Quality, Reliability, and Process Improvement

Quality, Reliability, and Process Improvement

Eighth Edition

Practice-Tested Methods and Procedures,
Based on Scientific Principles
and Simplified for Immediate Application
in a Variety of Manufacturing Plants

Norbert Lloyd Enrick, Ph.D.
Kent State University

INDUSTRIAL PRESS INC.
200 Madison Avenue
New York, New York 10157

Library of Congress Cataloging in Publication Data

Enrick, Norbert Lloyd, 1920–
   Quality, reliability, and process improvement.
   Rev. ed. of: Quality control and reliability. 7th ed. 1977.
   Includes bibliographical references and index.
   1. Quality control. 2. Reliability (Engineering)
I. Title.
TS156.E56   1984   658.5′62   85–17
ISBN 0-8311-1125-9

First Printing

QUALITY, RELIABILITY, AND PROCESS IMPROVEMENT—
Eighth Edition

Contents

v

Functional Table of Contents

CONTROL OF QUALITY

PRODUCT & PROCESS IMPROVEMENT

ASSURANCE OF RELIABILITY

MANAGEMENT ASPECTS

NOTE: An *ordinary* Table of Contents, reflecting the development of methods and concepts, from basic to more advanced, is designed to effectively and smoothly lead the reader and student into and through the material in the book. A *functional* table serves primarily as a *reference and review* of major categories of topics covered.

List of Tables

List of Tables *(Continued)*

Preface

The notion that quality control through statistical methods can save costs and enhance product marketability at the same time was originally conceived and developed by the great Dr. Walter A. Shewhart, in the late 1920s. His book *Economic Control of Quality of Manufactured Product* emphasized the cost saving aspects of a good quality control program. He described his work as "... an indication of the direction in which future developments may be expected to take place." But could he have foreseen the enormous impact of his creation?

During World War II, statistical quality control was endorsed and promoted by the Armed Services as a means of saving scarce material resources while at the same time seeking to ensure requisite quality of procurements. As a manufacturing methodology, and a means for efficient vendor-vendee relationships, the statistical and management aspects of the Shewhart approach and acceptance sampling were major contributors to the winning of that war. But there followed a period of decades when quality control methods received less attention than they deserved. It remained for Japanese industry—engineers, management, staff, and workers—to show the value of statistical quality control in terms of cost savings, productivity, and enhanced competitive position. As a consequence, quality has become a worldwide concern for manufacturers and marketers.

But it is not enough for management to desire good quality control, or to exhort people toward quality performance, or to set up quality control systems. Quality control methods will work for an organization only when installed carefully, knowledgeably, and competently. General information, desires, and wishes will not do anything. Those engaged in quality control, from top management to the operator on the shop floor or other production area, must become aware that quality starts with the *conception of the product*, as represented by design and development, and ends with the *consumer* and his or her long-term satisfaction with the product.

Modern quality control is founded on statistical principles. Sampling plans, derived from probability laws, serve to minimize inspection costs in relation to a desired level of quality protection. On control charts, statistically set limits warn when an operation is about to trend out of control. Statistical evaluation of tolerances and systematic investigation of processing variables help to obtain optimal conditions. Specially designed experiments, in the laboratory and in production, further aid in these optimization endeavors.

Some years ago, a book might have ended here. Today the growing interest of consumers in long-term product performance guarantees and the critical needs of high-technology products focus attention on the *time dimension of quality*, that is, reliability. For how long will a product perform well under specified usage and environmental conditions? Reliability-oriented design, evaluation of reliability, and assurance of reliability thus receive special attention.

While this book gives attention to management aspects, particularly the importance and application of participative methods in the areas of quality and productivity attainment, the principal emphasis is on *simplified but gap-free presentation of the important statistical methods* that form the heart of the successful program. Illustrations are given from quality and reliability improvements, productivity enhancement, and cost savings applications. No amount of good will can succeed without an understanding of how the "statistical gears" work in achieving product that is "made right the first time," thus saving on scrap and rework, simplifying processing sequences of operations, and reaching the market with an attractive, competitive package.

A Note to Teachers: Teaching aids and other materials are available for use in courses and seminars. Please contact the publisher.

A Note on English, Metric, and SI Units of Measurement

The world is turning to the metric system, and destined to become universal is the Système International (SI), which is a metric system with the kilogram (at some 2.2 pounds) as the basic unit of weight and the meter (approximately 3.3 feet) as the basic unit of dimension. Metric units have the advantage of ready conversion by shifting decimals or adding zeros. A millimeter is thus one-thousandth of a meter or 0.001 meter, while 1,000 meters are one kilometer. Within the English system, on the other hand, conversion involves division or multiplication by values other than 10. For example, 3 feet make a yard, 12 inches make a foot.

Illustrations in this book involve both English and metric units. In *all* instances, however, the examples are *not* designed to give absolutes but rather to demonstrate a principle. Thus, to show the method of obtaining an average, using values 50, 60, 45, 55, and 45 for tensile strength test results, we first note that the total is 250. Next, since 5 values were added, we divide by 5 to obtain the *average* 50. Further, a measure of variation is gained by subtracting the lowest value, 45, from the highest, 60, yielding the *range* 15. In performing these operations the units of measurement are of no consequence to us. They might be pounds or kilograms. What is important is the methodology—in this instance, the procedure for finding a sample average and range.

Consequently, regardless of a reader's preferences, and regardless of whether he understands only English or metric units, this detail will have no bearing on his ability to profit from *all* of the examples and applications to be presented.

CHAPTER 1

Inspection and Testing for Quality Control

Three major methods are available for the inspection and testing of products to control the quality of output produced. These are: (1) screening, (2) lot-by-lot inspection or sampling, and (3) process inspection. A more recent extended application of testing involves the assurance of product reliability. All of these matters will be discussed in the following chapters.

Screening

If every unit of product is inspected and the defective ones are screened out, this is called *screening* or *100-percent inspection*. Screening is often preferred to any other method of inspection because "only if you inspect every piece can you be sure to catch all the defectives." However, experience has shown that *100-percent inspection does not guarantee a perfect product*. Since the monotony of screening inevitably creates fatigue and lowers attention, no inspector can ever hope to catch all of the defectives. He always misses some and usually more than have been anticipated. If a perfect product is to be guaranteed, at least a 200-percent inspection is usually necessary unless some completely automated inspection device can be used.

In most types of mass production, screening can be used only sparingly because it is expensive and time consuming and interferes with the flow of work. Furthermore, there are many types of "destructive" testing where 100-percent inspection would result in 100-percent destruction of product. Examples are the sharpness testing of razor blades, which dulls the edges; tensile strength tests of wire, which snap the wire; chip tests of enamel, which crack the enamel; and firing tests of ammunition. With such products, unless a nondestructive test can be developed, sampling becomes a necessity.

Screening is generally the method of choice for vital parts upon which the functioning of an entire assembly depends, for parts produced by a process that normally turns out a relatively high proportion of defectives, and for outgoing products that must be inspected and tested for a high degree of quality assurance. Because routine 100-percent inspection soon becomes a part of regular production operations, it should be controlled by lot-by-lot sampling inspection, as will be discussed next.

Lot-by-Lot Inspection

Lot-by-lot inspection was invented to overcome the high cost of screening. A lot may consist of an accumulation of product, such as incoming material, partly finished articles and subassemblies, or completed product. Instead of examining each piece, the inspector limits himself to checking a relatively small number of sample pieces and then judges from them the acceptability of the whole lot.

The limitations of this procedure are that a sample does not always give a true picture of the entire lot from which it has been selected. For example, a lot may contain a great proportion of defectives, but since the inspector happens to pick up only good pieces in a small sample, he may erroneously accept the lot even though the lot as a whole is bad. On the other hand, a lot may contain only a few bad pieces, and yet these may be included in the inspector's sample, resulting in the unwarranted rejection of a good lot. These errors, known as *sampling errors*, are a familiar headache to experienced inspectors.

Here, modern statistics of mathematical probability help the inspector by furnishing him with ready-made sampling plans that guarantee a *minimum amount of inspection for a maximum amount of protection* against sampling errors. How to achieve this is the main topic of Chapter 2, "Installing Lot-by-Lot Inspection."

Without proper sampling plans, there usually will be either too much inspection or else not enough, resulting in excessive costs or ineffective control.

Process Inspection

Process inspection is accomplished by an inspector who patrols an assigned area, checking up on equipment, methods of operation, and occasional pieces of product from raw material to finished article. The purpose of process inspection is to discover defective products where and when they occur, so that corrective action can be taken immediately. Process inspection is concerned with all causes of defective work, be it operator, operation, equipment, or raw material.

A limitation of process inspection is that inspectors cannot be stationed at all machines at all times. As a result, considerable quantities of defective material may slip through between inspectors' visits. This applies especially to difficult control operations—such as precision machining, intricate casting, certain types of spot and resistance welding, and many kinds of chemical, mechanical, and electrical measurements. In such cases, the roving inspector often finds the faulty operation only after the damage in the form of defective work has already been done.

From this experience arose methods which *show quickly when something is wrong or is about to go wrong with the process*, sometimes even before defective work makes its appearance. The most important achievement of the science of statistical quality control is to have developed such methods and to have brought them into a form which can be used easily without difficult mathematics. These methods are known as the *control chart system*, which is the main topic of Chapters 4 through 6.

When Not to Install Statistical Quality Control

One does not merely install quality control in a plant. The matter must be considered thoroughly in order to decide upon a plan. Such preliminary thinking may indicate that statistical quality control is not applicable to the majority of operations in a given plant.

For example, if you are stamping out nonprecision metal blanks on punch presses, it will be sufficient for the operator or foreman to give the product an occasional spot check for burrs and scorch marks. This will be sufficient to detect deficiencies in the blanking die or material and insure a satisfactory product. Generally speaking, any elaborate quality control is uncalled for on product or operations where all the following conditions exist:

1. The product is a nonprecision product.
2. The quality of the product can be checked quickly.
3. Defective work is unlikely.

Reliability Testing

The age of space travel—automation and computers, sophisticated machinery and equipment, and intricate devices—has spotlighted an old problem and given it new urgency: to ascertain for how long a product may be expected to function properly; hence, how reliable it is. For a miniaturized circuit, for example, you will need to know the expected life. Next, if a hundred similar devices are working in an integrated manner in a computer, how long will the system operate in a trouble-free manner? As another illustration, consider a knitted synthetic heart valve. How many millions of cycles of trouble-free pumping—the life of the valve and of the patient—can we reasonably expect, given the environment in which the device is to operate? For answers to such questions, experts resort to reliability testing and evaluation, which in turn give rise to reliability design and redesign. Chapters 16 through 18 are devoted to problems of this nature and to ways and means of dealing with them.

Quality control within a plant can go a long way in assuring a product with proper performance characteristics. But we also need the additional testing, evaluation, and assurance work afforded by reliability engineering.

Systems Approach

Inspection operations within a manufacturing plant must be part of an integrated system. The various processes must be considered and decisions made on where to utilize screening, where to apply lot-by-lot inspection, and where to rely on control charts. Additionally, proper safeguards regarding acceptance of incoming materials and review of outgoing products are required. At each stage, the costs of inspection must be weighed against the benefits expected.

Maintenance of inspection records is vital for a well functioning quality control system. From a knowledge of the quality history of each operation, management will be in a position to determine the standards and tolerances that can normally be maintained. In turn, decisions on whether or not certain customers' requirements can be met, or whether equipment should be rebuilt or replaced, or whether adjustments are needed in product design, will often rely heavily on quality and related information accumulated from inspection and quality control records. When reliability of complex equipment is to be assured, availability of past performance data for valid assessments of expected product performance is indispensable.

With modern data processing techniques, the building up of essential, pertinent information, as part of an overall management system, is readily assured. Nevertheless, the usefulness of the information developed is no better than the inspection, quality control, and reliability assurance program from which the source data are derived.

REVIEW QUESTIONS

1. What are the three major methods of inspection and testing for quality control?
2. A production lot containing 20 percent of defectives is screened. Assume screening is 90 percent reliable (10 percent of the defectives are missed). In a lot of 1,000 pieces, how many defectives will remain after screening?
3. Assume that the production lot, just discussed, is subjected to 200 percent inspection. How many defectives will remain after these operations?
4. Give several examples of (A) destructive and (B) nondestructive testing.
5. Sampling errors are not "mistakes." What, then, produces the phenomenon of sampling error?
6. In what manner is reliability testing distinct from quality control testing?
7. What purposes are accomplished by use of the systems approach in the installation and maintenance of a quality control program?

CHAPTER 2

Installing
Lot-by-Lot Inspection

Four steps are to be followed in setting up lot-by-lot inspection for a particular item:

1. Set up inspection lots
2. Arrange for rational lots
3. Establish an acceptable quality level
4. Select a sampling plan.

Step 1. Set Up Inspection Lots

Under lot-by-lot inspection, the size of the inspection lot may vary from about 300 articles up to any number. For smaller lots, control charts or special sampling plans will be preferable.

As a rule the quantity of product which normally moves through the shop in a single lot is also your inspection lot. For example, a hand-truck with two tote boxes, each containing 300 blanks, may be a practical lot in the press-working department. On the other hand, a barrel of 5,000 galvanized washers might be a practical lot in the plating department.

Theoretically, there is no upper limit to the size of the lot. But in practice lots should be kept small enough so that they are easy to move and do not require special handling.

Another factor influencing lot size is how frequently work from one machine is to be inspected. For example, a hammer is producing approximately 1,000 multiple die drop forgings per hour. You wish to inspect the work every half hour so that if quality should go out of control, you can detect it fairly soon; therefore the lot size should be approximately 500 each. In other words, small lots give you a better control over the process and prevent waste of materials; but it is not desirable to make lots *too* small. As a rule, 300 pieces is the minimum size as previously mentioned. If production lots are smaller than 300, it is better to accumulate two or three of them until you have a sizable inspection lot. You do not save much by sampling lots smaller than 300 and might as well do 100 percent inspection on them, unless control charts (to be discussed later) can be used as an alternative.

Step 2. Arrange for Rational Lots

A "rational lot" is one whose units have been produced from the same source. As much as possible, a lot should consist of articles produced from one batch of raw materials; one production line or machine; one mold, pattern, or die; and one and the same shift. Often it is not possible in practice to separate product as strictly as that, but you should adhere to the rule of forming rational lots as closely as possible. If you mix up products from different sources and find a bad lot, you cannot put your finger immediately upon the source of trouble—which can be done if there are rational lots.

To illustrate, a die-casting department has three different tumbling barrels for washing and polishing small zinc castings. Lots are kept separate, and a note on each route tag indicates in which barrel each lot has been tumbled. Now if the inspector finds a lot which contains too many chipped, broken, or poorly polished castings, he can immediately locate the barrel in which the trouble occurred by merely referring to the route tag. On the other hand, if lots are mixed up, it will take much longer to find out which tumbling barrel is not working properly.

Step 3. Establish an Acceptable Quality Level

The idea of an acceptable quality level is arrived at from the following consideration: under the speed of mass production, it is often impossible to turn out continually 100 percent satisfactory product. One must assume that a certain proportion of defectives will always occur on certain processes. However, if the percentage does not exceed a certain limit, it is often more economical to allow the defectives to go through rather than to screen each lot. This limit is called the *acceptable quality level* (AQL).

To illustrate, in a certain automatic polishing operation it was found that up to 1 percent of defective pieces (incompletely polished) was not abnormal. Since it would be rather expensive to screen each lot after polishing, it was decided to consider 1-percent defective product as the AQL, even though this small quality of bad material would go through to plating and finishing only to be thrown out in the final screening before shipment.

To give a second illustration: in a certain centrifugal casting operation, it was found that up to 2 percent of the work might normally be expected to be defective. Because of the high cost of further processing operations on this casting, it was determined that a 2-percent level of defectives was not allowable. This meant that screening had to be installed.

Thus the principle of establishing the acceptable quality level becomes apparent: As soon as one has determined the normal percentage defective which the particular operation produces, one has to ponder the important question: Is the normal percent defective produced *acceptable* or *not*? If you have decided that it *is accepta-ble*, then you should use a proper sampling plan to see that, in the *long run*, no more than this allowable proportion of defectives is passed even though some accepted lots have more than this. If you have decided that the normal percentage defective which the machine produces is *not acceptable*, then you should install screening inspection on every lot. By doing so, you achieve two important objectives: You assure that normally no more than the allowable percentage of defectives goes through further production, and you also know when something goes wrong with the process.

One question still needs to be answered: *How do we establish the normal percent defective?* To do this we have to watch the product for some time while it is under regular operation. Ordinarily, sufficient data can be accumulated over a period of one to two weeks to determine the percentage of defectives normally occurring in the product from certain equipment and operations. Already existing records of past performance are also a good guide in determining the capability of the machine.

To illustrate, records accumulated from inspection of product revealed the following percentages of defectives produced in one day on three 500-pound drop hammers:

	Hammer No. 1	Hammer No. 2	Hammer No. 3
Lot No. 1	2.0%	0.8%	1.9%
Lot No. 2	1.8%	1.3%	2.2%
Lot No. 3	1.3%	7.4%*	2.5%
Lot No. 4	1.1%	0.5%	0.8%
Lot No. 5	1.3%	1.2%	1.2%
Lot No. 6	1.8%	1.2%	1.2%
Lot No. 7	0.9%	1.4%	0.8%
Lot No. 8	4.7%*	0.9%	2.6%

The two percentages marked with an asterisk represent lots containing an unusually high portion of defectives. Investigation revealed faulty dies and failure to correct the set-up promptly as the causes of the poor quality. The percentages observed here were considered clearly abnormal. Most of the remainder of the lots showed defectives varying roughly between 1 and 2 percent, which could be considered normal. A few figures were somewhat above 2 percent, but in the absence of any demonstrable mechanical failure of the equipment, carelessness of operators had to be assumed as responsible.

Data, such as given above, were collected for several additional days to make sure that all normal operating conditions would be covered. Comparisons were made to analyze differences in scrap due to different operators and different dies. The final conclusion reached was that up to 2 percent defectives in any lot should be considered as not abnormal, and this figure was finally adopted as the acceptable quality level.

The example presented indicates that a certain amount of judgment, rather than precise scientific measurements, will enter into determinations of the acceptable quality level. Sometimes asked is the question of whether there are any standard tables furnishing definite values of the normal percent defective for various machines and operations. The answer is *no*. To initiate such a program of standard tables seems hopeless because account would have to be taken of such variables as age and condition of all types of equipment, differences in raw materials, skill of operators, degree of maintenance, and many other factors. Therefore, it is still necessary to make individual determinations in each shop. Before any final decision is made on the choice of an acceptable quality level, the inspection supervisor should consult the engineering, planning, and cost departments of his plant.

As regards *incoming materials*, the applicable value of the AQL will depend upon commercial standards or the firm's own specifications. Incoming materials may be sampled statistically just as work-in-process or end product. Accordingly, some technical committees of manufacturers' associations have established recommended standards of AQL in terms of the percentage of defective product in a shipped lot. Individual large buyers often specify the AQL and the method of sampling of incoming materials as part of the purchasing contract or purchasing order. Prominent in this field are the procuring departments of various agencies of the United

States government, department stores, chain stores, and various manufacturing organizations. Going one step further, the purchaser may require the supplier to submit evidence that his inspection and testing, sampling and quality control are such as to permit attainment of acceptable quality levels.

Step 4. Select a Sampling Plan

Let us assume that the acceptable quality level on a certain item is established at 2 percent and that inspection lots consist of approximately 750 pieces each. If you want to make sure that in the long run the average of defectives passed will not exceed 2 percent, you have to know:

1. How many sample pieces to inspect in each lot
2. When to accept the lot
3. When to reject the lot

This is simple. All the information is in ready-made sampling plans, such as are supplied by the accompanying master sampling table (Table 2-1)[1].

Example 1: To illustrate, let us assume that the acceptable quality level for a certain metal stamping is 2 percent. Lot sizes consist of approximately 750 pieces each. To find the applicable sampling plan, the following method is used:

Step 1: The master table is entered under "lot size" for lots of 500 to 799. In the second column of this horizontal section are found the sample sizes: 40, 60, 80, 100, 120.

Step 2: Moving to the right, in the same horizontal section, to the column for 2 percent, the acceptance and rejection numbers are obtained; thus for *A:* 0, 1, 1, 2, and 4; and for *R:* 3, 4, 5, 5, and 5.

Step 3: By combining these figures, the complete sampling plan is obtained:

Sample Size	Acceptance Number	Rejection Number
40	0	3
60	1	4
80	1	5
100	2	5
120	4	5

Now you start using the plan. Begin by selecting 40 sample pieces at random from different parts of the lot. If you find no defectives,

---

[1]The statistical basis of the sampling plans tabulated here is explained in the next chapter.

Table 2-1.　Master Sampling Table

ACCEPTABLE QUALITY LEVEL (column headings below)

LOT SIZE	SAMPLE SIZE	0.25		0.5		0.75		1		1.5		2		3		4		5		6		7		8		9		10		12	
		A	R	A	R	A	R	A	R	A	R	A	R	A	R	A	R	A	R	A	R	A	R	A	R	A	R	A	R	A	R
499 or less	40	→		→		→		0	2	0	2	0	3	1	4	1	4	1	6	2	6	2	7	3	7	3	8	4	9	4	9
	50	→		→		→		0	2	0	3	1	3	1	4	2	5	2	6	3	7	3	8	4	9	4	9	5	10	5	11
	60	→		→		→		0	3	1	3	1	3	2	5	3	6	3	7	4	8	4	9	5	10	5	11	7	12	7	13
	70	→		→		→		1	3	1	3	2	4	2	5	3	6	4	8	5	9	5	9	6	10	7	11	8	13	8	13
	80	→		→		→		2	3	3	4	3	4	4	6	5	6	7	8	8	9	8	9	9	10	10	11	12	13	12	13
500 to 799	40	→		→		*	1	*	2	0	3	0	3	0	4	1	5	1	5	1	6	1	7	2	8	2	8	2	9	4	10
	60	→		→		*	1	0	2	0	3	1	4	1	5	2	6	2	7	3	8	3	9	4	10	5	11	5	11	6	12
	80	→		→		0	2	1	3	1	4	1	4	2	6	3	7	3	8	5	10	5	11	6	12	7	13	8	14	9	15
	100	→		→		0	2	1	3	2	4	2	5	3	6	4	8	5	9	6	11	7	13	8	13	9	16	10	17	12	18
	120	→		→		1	2	2	3	3	4	4	5	5	6	7	8	8	9	10	11	12	13	13	14	15	16	16	17	18	19
800 to 1,299	40	→		*	1	*	1	*	2	*	3	0	3	0	4	0	5	0	6	1	6	1	7	1	8	2	8	2	9	2	10
	60	→		*	1	0	2	0	2	0	3	0	4	1	5	2	6	2	7	2	8	3	9	3	10	4	11	4	12	5	12
	80	→		0	2	0	2	0	3	1	4	1	5	2	6	3	7	3	8	4	10	5	11	5	12	6	13	8	15	8	15
	100	→		0	2	1	2	0	3	1	4	2	5	3	6	5	8	5	10	5	11	7	13	7	14	9	15	10	17	10	18
	120	→		0	2	1	3	1	3	2	5	3	6	3	7	6	9	6	10	7	13	8	14	9	16	11	18	12	19	13	21
	160	→		1	2	2	3	3	4	4	5	5	6	7	8	9	10	10	11	13	14	15	16	16	17	18	19	19	20	22	23
1,300 to 3,199	50	*	1	*	1	*	2	*	3	*	3	0	4	0	4	0	5	0	6	1	7	1	8	2	9	2	10	3	10	3	11
	75	*	1	*	1	0	2	0	3	0	4	0	5	1	5	2	7	2	8	3	9	4	10	4	12	5	12	6	14	6	15
	100	0	1	0	2	0	3	1	4	1	4	1	5	2	6	4	8	4	9	5	11	6	12	6	14	8	15	9	17	10	18
	125	0	2	0	2	1	3	1	4	2	5	2	6	3	7	6	9	5	11	7	13	8	15	9	16	11	18	12	20	13	21
	150	1	2	1	2	2	3	2	5	2	5	3	7	4	8	8	10	7	13	9	15	10	17	11	19	14	21	15	23	16	25
	200	2	3	1	2	3	4	4	5	5	6	6	7	8	9	10	11	13	14	17	18	17	18	20	21	22	23	25	26	27	28

A—Acceptance number; R—Rejection number; *No acceptance at this sample size.
Arrows: When there is an arrow under a given AQL, use the first sampling data below arrow. (Form larger lots if possible.)

Table 2-1 (*Continued*).   Master Sampling Table

LOT SIZE	SAMPLE SIZE	0.25		0.5		0.75		1		1.5		2		3		4		5		6		7		8		9		10		12	
		A	R	A	R	A	R	A	R	A	R	A	R	A	R	A	R	A	R	A	R	A	R	A	R	A	R	A	R	A	R
3,200 to 7,999	50	*	1	*	2	*	2	*	3	*	3	*	4	*	5	0	6	0	7	0	8	1	9	1	10	1	11	2	11	2	12
	100	*	1	*	3	0	3	0	4	1	4	1	5	1	7	2	8	3	10	4	12	5	13	5	15	6	16	8	17	9	19
	150	0	2	0	3	1	4	1	5	2	5	2	6	2	8	5	11	5	13	8	15	9	17	10	19	11	21	13	23	15	25
	200	0	2	1	3	2	4	2	5	3	6	3	7	4	10	7	13	8	16	12	18	13	21	15	24	16	26	19	29	21	31
	250	0	2	1	4	2	5	2	6	4	8	5	8	6	11	9	15	11	18	15	22	17	25	19	28	22	32	25	35	28	38
	300	1	2	3	4	4	5	5	6	7	8	8	9	10	11	15	16	17	18	22	23	25	26	29	30	32	33	36	37	39	40
8,000 to 21,999	100	*	2	*	3	*	3	*	4	0	5	0	6	0	7	1	9	1	11	3	12	3	14	4	16	5	17	↓	↓	↓	↓
	150	*	2	*	3	0	4	0	5	1	6	1	7	2	9	3	11	4	14	6	16	7	17	9	20	10	22	↓	↓	↓	↓
	200	0	2	0	4	1	5	1	6	2	7	3	8	3	10	6	13	7	17	10	19	11	22	13	24	16	27	↓	↓	↓	↓
	250	0	2	1	4	2	5	2	6	2	8	4	9	5	12	8	16	10	19	13	23	15	25	18	29	19	32	↓	↓	↓	↓
	300	0	3	1	4	2	6	2	7	3	8	5	11	6	13	10	18	12	22	16	26	19	29	22	33	24	37	↓	↓	↓	↓
	400	1	3	3	6	3	7	4	8	5	10	8	14	10	16	14	22	18	27	23	33	27	37	31	42	34	47	↓	↓	↓	↓
	500	2	3	4	7	5	7	7	8	9	10	13	14	16	17	22	23	28	29	34	35	40	41	45	46	50	51	↓	↓	↓	↓
22,000 to 99,999	100	*	2	*	3	*	4	*	5	*	6	*	8	0	8	0	11	1	13	2	16	↓	↓	↓	↓	↓	↓	↓	↓	↓	↓
	200	0	3	0	4	0	5	0	6	1	7	1	10	3	11	3	15	6	18	8	20	↓	↓	↓	↓	↓	↓	↓	↓	↓	↓
	300	0	3	1	4	1	6	1	8	2	9	3	12	5	14	7	19	11	23	14	26	↓	↓	↓	↓	↓	↓	↓	↓	↓	↓
	400	1	4	2	5	2	7	3	9	4	11	5	14	8	17	11	23	16	28	21	33	↓	↓	↓	↓	↓	↓	↓	↓	↓	↓
	600	1	4	3	6	3	9	5	11	7	14	10	18	14	23	19	31	27	38	34	46	↓	↓	↓	↓	↓	↓	↓	↓	↓	↓
	800	1	4	3	7	5	11	8	14	11	17	14	23	20	28	27	38	37	49	46	58	↓	↓	↓	↓	↓	↓	↓	↓	↓	↓
	1000	3	4	6	7	10	11	13	14	17	18	22	23	30	31	40	41	53	54	65	66	↓	↓	↓	↓	↓	↓	↓	↓	↓	↓
100,000 and up	200	*	3	*	6	*	7	*	8	0	9	0	13	2	13	2	18	↓	↓	↓	↓	↓	↓	↓	↓	↓	↓	↓	↓	↓	↓
	400	0	4	0	7	0	8	0	10	3	12	3	17	7	18	9	25	↓	↓	↓	↓	↓	↓	↓	↓	↓	↓	↓	↓	↓	↓
	600	1	5	1	8	2	10	2	12	5	15	7	20	12	22	17	33	↓	↓	↓	↓	↓	↓	↓	↓	↓	↓	↓	↓	↓	↓
	800	2	5	2	9	4	11	5	14	8	18	11	24	18	29	24	40	↓	↓	↓	↓	↓	↓	↓	↓	↓	↓	↓	↓	↓	↓
	1000	2	5	3	9	5	13	7	17	11	21	15	28	23	35	32	47	↓	↓	↓	↓	↓	↓	↓	↓	↓	↓	↓	↓	↓	↓
	1200	4	5	5	9	7	15	9	19	14	24	19	32	28	40	39	55	↓	↓	↓	↓	↓	↓	↓	↓	↓	↓	↓	↓	↓	↓
	1600	4	5	8	9	14	15	19	20	24	25	33	34	45	46	61	62	↓	↓	↓	↓	↓	↓	↓	↓	↓	↓	↓	↓	↓	↓

ACCEPTABLE QUALITY LEVEL

Arrows: When there is an arrow under a given AQL, use the first sampling data above arrow.

accept the lot since the acceptance number is 0. If you find 3 or more defectives, reject the lot, since the rejection number is 3. But what if you find 1 or 2 defectives? The defectives you have found lie between the acceptance and rejection numbers, and you can neither accept nor reject the lot. The thing to do then is to gather additional evidence by taking another sample of 20 from the lot. This brings the total sample size up to 60 (40 plus 20). Again compare the total number of defectives found up to this point to determine whether to accept or reject the lot or to continue sampling. This process may go on until the highest sample size—120—is reached, at which point the gap between the acceptance and rejection numbers disappears and the problem of acceptability is finally solved.

It should be understood that this sampling table has been set up on the basis of an *overall* percent defective—that is, the final percent of defective parts that can be expected over a considerable run. This table therefore rests on the assumption that all rejected lots will be screened so that they become theoretically perfect and thus will raise the general average quality of the lots passing through.

In the particular case covered by Example 1, for instance, the acceptance number of 4 in a sample of 120 actually represents a maximum of 4/120 or 3.3 percent defective, whereas the AQL allows 2 percent. Thus, individual lots having between 2 and 3.3 percent defectives will occasionally be passed through (not considering those bad lots which slip by now and then because of the "luck of the draw" errors normally present in the sampling process), but in the long run the overall proportion of defectives will be held at 2 percent. The probability of occasional errors of rejecting good lots and accepting bad ones is discussed in the next chapter, with specific reference to the sampling plans given in Table 2-1.

Example 2. Let us assume that for another item usual lots sizes are approximately 3,000, and the acceptable quality level is 1 percent. Then the following sampling plan is obtained from the master table by referring under "lot size" to lots ranging from 1,300 to 3,199.

Sample Size	Acceptance Number	Rejection Number
50	*	3
75	0	3
100	1	4
125	1	4
150	2	5
200	4	5

This particular plan differs from the one previously presented chiefly in the fact that it contains an asterisk in place of the first acceptance number. As indicated in the master sampling table, an asterisk means no acceptance can be made at this sample size. In other words, you can reject a lot if you find three or more defectives in your first sample, but you cannot accept it even if you find no defectives at all, because the sample size is too small. In the latter case, you will have to examine an additional 25 articles, bringing your total sample size up to 75 (50 + 25). This new sample is sufficiently large to permit acceptance of the lot if no defectives have been found. In all other respects, you use this plan the same way as the one in the preceding illustration.

The Least Amount of Inspection

The type of sampling just described is called *sequential sampling* because a sequence of samples is taken. You may ask: "Why bother with a sequence of samples? Are there no plans that require only one sample of definite size to make a decision?" There are such plans, and they are known as *single sampling plans*. In such cases a master table of single sampling plans (Table 2-2) is used. These plans are not so economical because sequential sampling usually requires much less inspection to determine the quality of any lot than single sampling plans do.

At first glance, this statement may seem paradoxical, because in Example 1 we might have to inspect 120 pieces, but this is not likely to happen very often. As a rule, a sample of 40 or 60, under the conditions of Example 1, would give the desired answer. If it occasionally happens that you have to do much additional sampling on any lot, this only proves that the lot was on the borderline between good and bad and really needed more sampling to arrive at the right decision. If a lot is *very good* or *very bad*, then usually the inspection of the first sample will give you the proper answer. In general, sequential sampling costs half as much as single sampling.

A special type of sequential sampling is *double sampling*, developed by H. G. Romig and H. F. Dodge of the Bell Telephone Organization, and used widely throughout the Bell System and Western Electric. Instead of providing for several sequential samples to be inspected, double sampling plans require only *two* samples. Since the recently developed multiple sequential sampling plans save considerably more inspection than double sampling, double sampling plans have not been included in this treatise.

Single Sampling Versus
Sequential Sampling

Despite the fact that sequential sampling requires, on the average, only half as much inspection as single sampling, there are some plants where single sampling is still preferred. The reason is one connected with material handling problems. If the product is bulky and difficult to move, such as large castings, heavy coils, etc., it will be less costly to move a large number of sample articles to a place of inspection at *one time*, rather than to move a much smaller number of units *several times*.

How to Use Single Sampling Plans

The application of single sampling is best explained by an example. Let us assume that the acceptable quality level for a certain item is 1 percent. Each lot consists of approximately 300 units. From the master table of single sampling plans (Table 2-2) the individual sampling plan needed is selected as follows:

Step 1: The master table is entered in the horizontal section for lots of 499 or less. In the second column of this section is found the sample size 75.

Step 2: Moving to the right in the same horizontal section to the column for 1 percent, the acceptance and rejection numbers are found:

$$A \qquad\qquad R$$
$$1 \qquad\qquad 2$$

Step 3: By combining these figures, the complete sampling plan is obtained:

Sample Size	Acceptance Number	Rejection Number
75	1	2

The plan is now ready for use. You simply inspect a sample of 75 pieces. If not more than one defective piece is found, accept the lot, since the acceptance number is 1. If you find two or more defectives, reject the lot, since the rejection number is 2.

For comparison, the particular sequential sampling plan corresponding to the single sampling plan is furnished below. This comparison shows that while single sampling plans are simpler to apply, sequential sampling plans require much fewer articles to be

Table 2-2.   Master Table of Single Sampling Plans

LOT SIZE	SAMPLE SIZE	ACCEPTABLE QUALITY LEVEL																					
		0.25		0.5		0.75		1		2		3		4		5		6-7		8		9 to 12	
		A	R	A	R	A	R	A	R	A	R	A	R	A	R	A	R	A	R	A	R	A	R
499 or less	75	→		→		→		1	2	2	3	3	4	4	5	5	6	6	7	8	9	10	11
500 to 799	115	→		→		→		2	3	3	4	4	5	6	7	8	9	9	10	11	12	14	15
800 to 1299	150	→		→		1	2	3	4	4	5	5	6	8	9	10	11	12	13	15	16	20	21
1300 to 3199	225	→		1	2	2	3	4	5	5	6	8	9	11	12	14	15	17	18	21	22	29	30
3200 to 7999	300	→		2	3	3	4	5	6	7	8	10	11	14	15	18	19	22	23	26	27	←	
8000 to 21999	450	1	2	3	4	4	5	6	7	9	10	14	15	20	21	26	27	←		←			
22000 to 99999	750	2	3	4	5	5	6	8	9	13	14	23	24	←		←							
100,000 and up	1500	3	4	6	7	8	9	13	14	23	24	←											

A—Acceptance number; R—Rejection number.

Arrows: When there is an arrow below given AQL, use first sampling data below arrows in upper left-hand corner (forming larger lots if possible) or first sampling data above arrows in lower right-hand corner, as case may be.

inspected on the average; in fact, a sample of 40 frequently will give the desired result. Again assuming a lot of 300 units and an AQL of 1 percent, we find in Table 2-1 the following:

Sample	Acceptance Number	Rejection Number
40	0	2
50	0	2
60	0	3
70	1	3
80	2	3

Meaning of Arrows in Master Sampling Tables

Acceptance and rejection numbers are not supplied in the upper left and lower right corners of the master tables. Instead, these sections contain arrows pointing downward and upward respectively. The downward arrows indicate that sample sizes adjacent to arrows are not large enough and larger samples should be employed; the upward arrows indicate that smaller samples should be employed. This is illustrated by the following example:

Example 1: For a certain item assume that the AQL is 0.5 percent and the usual lot size is approximately 1,000. Find a single sampling plan for this item.

Procedure: Examination of Table 2-2 discloses that no acceptance and rejection data are available for lots of 800 to 1,299 when the acceptable quality level is 0.5 or less; instead there is an arrow pointing downward. This means that the next lower section of the table must be used. This is the section under 0.5 for lots of 1,300 to 3,199, which is reproduced below:

Sample Size	Acceptance Number	Rejection Number
225	1	2

The procedure now is the same as described previously. Sequential plans are established in a similar manner.

Example 2: Assuming that the lot size is 30,000 and the AQL 7 percent, find a sequential sampling plan.

Procedure: Refer to Table 2-1 and the section for lot sizes of 22,000 to 99,999. An arrow under 7 pointing upward indicates that the data for lot sizes 8,000 to 21,999 should be employed.

Random Sampling

Samples must be selected in a truly random manner from different parts of a lot. This means that you should not examine locally. For example, do not concentrate on the top layer or the side layer of a lot, but take samples in such a way that each unit in the lot has an equal chance of being chosen. In the long run this is the only method likely to minimize misleading results. Hunting for defectives in the bottom corner of lots or any other such "system" is not only ineffective but also defeats the primary purpose of sampling—to obtain a true representation of the lot as a whole.

Major and Minor Defects

In special cases it may be found desirable to use two different sampling plans for one item—one plan to control major defects and another plan to control minor defects. The latter plan would have a higher AQL simply because the defects are of less importance.

To illustrate, in a particular type of stainless steel blanks, scale marks, die marks, and thick lines might be considered major defects with an AQL of 0.5 percent. At the same time, slight scratches, roll marks, light burrs, etc., might be considered minor defects with an AQL of 5 percent. Then the respective sampling plans are selected from the master table and used separately to accept and reject lots either on the basis of major or minor defects.

Inspection Records

Once the sampling routine has been established, it is desirable that the inspector keep a lot-by-lot record of the lots inspected. Examples of such records are given in Figs. 2-1 and 2-2.

Inspection records should be submitted at the end of each day to the supervisor of inspection. Properly summarized, they will give him a clear overall picture of the daily manufacturing operations and enable him to put his finger on sore spots and advise the production, engineering, and methods departments accordingly. In many plants, the standards department will use figures of inspection for establishing quality-quantity bonus systems.

Analysis of Entries Made on
Lot-by-Lot Inspection Record

On the record form presented in Fig. 2-1 entries are shown covering lot-by-lot inspection for major and minor defects in six lots of

LOT-BY-LOT INSPECTION RECORD

1. ITEM: S.S. Part — Dwg. # 20999
2. LAST OPERATION: Blanking
3. OPERATOR: # 35

DATE OF INSPECTION	1/1	1/1	1/1	1/2	1/2	1/2
LOT NUMBER	1001	1002	1003	1004	1002R	1005
LOT SIZE	400	399	401	500	360	400
ORIGINAL OR RESUBMITTED LOT (Check One)	✓ ORIG.	✓ ORIG.	✓ ORIG.	✓ ORIG.	✓ RESUB.	✓ ORIG.

MAJOR DEFECTS

SAMPLING TABLE FOR MAJORS / MAJOR DEFECTIVES

SAMPLE SIZE	ACC. NO.	REJ. NO.	NUMBER	NUMBER	NUMBER	NUMBER	NUMBER	NUMBER
40	0	2	1	3 x	0 ✓	0	0	0
50	0	2	1					
60	0	3	1					
70	1	3	1 ✓					
80	2	3						

TYPES OF MAJOR DEFECTS FOUND / MAJOR DEFECTS

	NUMBER	NUMBER	NUMBER	NUMBER	NUMBER	NUMBER
Scale marks	1					
die-mark		III				
thick lines		1				
pin holes		1				

MINOR DEFECTS

SAMPLING TABLE FOR MINORS / MINOR DEFECTIVES

SAMPLE SIZE	ACC. NO.	REJ. NO.	NUMBER	NUMBER	NUMBER	NUMBER	NUMBER	NUMBER
40	1	4	2	2	1 ✓	2	1	4 x
50	1	4	2			2		
60	2	5	2 ✓			3		
70	2	5				4		
80	4	5				4		

TYPES OF MINOR DEFECTS FOUND / MINOR DEFECTS

	NUMBER	NUMBER	NUMBER	NUMBER	NUMBER	NUMBER
Slight scratches	1		1	IIII		1
roll marks						
burrs		II				
small pin holes (superficial)					1	III

LOT ACCEPTED OR LOT REJECTED (Check One)	✓ ACC.	ACC. ✗ REJ.	✓ ACC.	✓ ACC.	✓ ACC.	ACC. ✗ REJ.

SIGNATURE: J. Jones

Fig. 2-1. Lot-by-lot inspection record showing major and minor defects after blanking operation.

LOT-BY-LOT INSPECTION RECORD

1. ITEM: *Cold Drawn Steel Bars* # 134 , 1½" diam.

3. OPERATOR: *J. Jones*

2. LAST OPERATION: *Temper*

4. DATE OF INSPECTION	4/1	4/2	4/2	4/3	4/3	4/4
5. LOT NUMBER	H 114	H 115	H 116	H 117	H 118	H 119
6. LOT SIZE (NO. OF PCS. IN LOT)	c. 450	c. 450	485	500	450	480
7. ORIGINAL OR RESUBMITTED LOT (CHECK ONE)	✓ ORIG. / RESUB.	✓ ORIG. / RESUB.	✓ ORIG. / RESUB.	✓ ORIG. / RESUB.	✓ ORIG. / RESUB.	✓ ORIG. / RESUB.

8. SAMPLING TABLE			9. MAJOR DEFECTIVES					
SAMPLE SIZE	ACC. NO.	REJ. NO.	NUMBER	NUMBER	NUMBER	NUMBER	NUMBER	NUMBER
40	0	2	0	1	4	0	0	0
50	0	2		1				
60	0	3		1				
70	1	3		1				
80	2	3		2				
CHECK NO.								

10. TYPES OF DEFECTS FOUND	11. DEFECTS					
	NUMBER	NUMBER	NUMBER	NUMBER	NUMBER	NUMBER
Handling marks		1	1			
scale			11			
laps			1			
bent		1				

12. LOT ACCEPTED OR LOT REJECTED (CHECK ONE)	✓ ACC. / REJ.	✓ ACC. / REJ.	ACC. / ✓ REJ.	✓ ACC. / REJ.	✓ ACC. / REJ.	✓ ACC. / REJ.
13. INSPECTOR'S SIGNATURE:			*K. Smith*			

Fig. 2-2.   Lot-by-lot inspection record of cold-drawn steel bars.

small stainless steel parts following blanking operation. The particular entries made for each lot are analyzed in the following:

Lot No. 1,001: The first sample of 40 blanks inspected showed 1 major defective (scale marks) and 2 minor defectives (1 scratch and 1 mark from the rollers feeding the steel coil into the blanking press). Scale marks tend to indicate improper processing, particularly in respect to pickling at the steel mill. However, in successive samples no additional defectives were found, and it was assumed that the one piece represented merely an isolated streak at the beginning of the coil. When 60 sample pieces had been inspected, the number of minor defectives found up to that point was still only 2, which was equal to the acceptance number for "minors." Therefore, the lot was accepted for minor defectives, but the lot could not yet be accepted for major defectives, since the number of defectives found up to this point was 1 and the acceptance number 0. Therefore the inspector had to inspect an additional sample of 10, bringing the total sample size up to 70. Since no additional defectives were found, and the new acceptance number was 1, the lot could be accepted for "majors." Thus, the lot passed the standards for both major and minor defects.

Lot No. 1,002: This lot revealed three major defectives in the first sample of 40, in addition to two minor defectives. Since for sample size 40 the acceptance number is 2 and the defectives found were 3, the lot was rejected. The die, which had begun to throw burrs and die marks, was sent to the toolroom for sharpening.

Lot No. 1,003: This lot was produced on another die. It could be accepted for both majors and minors on the first sample size.

Lot No. 1,004: This lot was accepted for majors on the first sample size; however, several minor defects were found, which called for additional inspection. The lot was accepted for minors only after a total of 80 sample pieces had been examined.

Lot No. 1,002R: Lot No. 1,002 had been rejected. Now it was resubmitted to the inspector as Lot 1,002R, after having been screened. It was accepted on the first sample size.

Lot No. 1,005: This lot was rejected on the first sample size because the sample showed it contained too many minor defects in the form of scratches and pinholes. The existence of the great number of pinholes led to a further investigation and rejection of three coils of steel from a particular shipment, since similar surface imperfections were evident.

It should be noted that in actual practice the inspector will rarely encounter the variety of defects shown in the specimen inspection record discussed here. These have been purposely increased in order to demonstrate all possible types of situations that may arise in the systematic application of sampling plans. As soon as a lot has been rejected, the foreman should be notified so that he can check into the trouble and correct the faulty operator, operation, or equipment that may be responsible for the bad lot. Thus, lot-by-lot inspection renders a most important service to the shop: it immediately arrests the flow of bad product and permits remedy before materials and time are needlessly wasted.

Disposition of Defective Material

If incoming materials are inspected and the lot is found to be unsatisfactory, its disposition is a matter of special concern to the purchasing department and production superintendent. If a lot consisting of work in process or completed product is rejected, it may be submitted to screening to take out all the defectives, or it may be sent directly to repair and salvage, sold as seconds, or be scrapped entirely, whichever procedure involves the least loss.

Defense Department Plans

A special set of single, double, and sequential sampling plans is provided in Military Standard 105, published for the Department of Defense by the Government Printing Office. These plans are more elaborate than those presented here, but in general no new principles are involved. Before using the government sampling tables, one should have an understanding of the statistical foundations of sampling systems, such as provided in the next chapter.

Sampling Continuous Product

In contrast to individual units of product, continuous items of product consist of uninterrupted lengths of material. Examples are belting, cloth, wire, cord, metal coil, and long tubing. These products often require a considerable amount of visual examination for defects. In the absence of individual sample pieces, an arbitrary length, such as 1 yard, is taken as the equivalent of a unit of product in applying sampling tables. However, here the difficulty is that it is impractical to select *at random* individual yardage sections from different parts of a roll. Therefore, the following compromise procedure is typical of what is often indicated:

1. Inspect continuous lengths of 50 to 75 yards each time.
2. Select these lengths from different parts of the lot.

The lot will consist of several rolls, of course.

Example: A lot consists of 15 rolls of wire strand, each 500 yards long. The total lot size is therefore 7,500 yards. This calls for a sample size of 525 yards as shown by Table 2-3. While the table gives sample sizes ranging from 525 to 1,275 for lot sizes of 9,999 or less, inspection begins with the smallest sample size, as previously explained.

Select one sample roll at random from the lot and inspect a continuous length of 75 yards near the beginning of the roll and 75 yards in about the middle of the roll. Then select another sample roll. This time inspect 75 yards about ⅓ way up the roll and 75 yards near the end of the roll. This process continues until the entire 525 sample yards have been inspected.

Special Sampling Tables

In the procedure just given, it is evident that *the principle of random sampling is not completely adhered to.* Consequently the usual sampling tables are no longer adequate, because they are constructed on the assumption that each individual sample piece has been taken *at random* from different parts of the lot.

To compensate for the deficiency in random sampling of continuous items of product, the sample size has to be enlarged: how much is shown in Table 2-3 for sequential sampling and in Table 2-4 for single sampling. The method of establishing individual sampling plans from these tables is similar in principle to that given for the prior tables in this chapter.

Long-Term Value of Lot-by-Lot Inspection

Application of sampling plans on a lot-by-lot basis assures that each successive lot undergoes a specific type of check as determined by management. As in most types of control activity, the prime value of the approach is not that checks are applied to an *individual* lot, but rather that a system of checking has been installed for a particular production operation. When management has been assured that all important processes are under effective quality surveillance, a major goal of overall control of operations will have been accomplished.

Records accumulated from lot-by-lot inspection, moreover, give valuable information regarding the general level of performance of a manufacturing operation. In one plant, for example, lot-by-lot inspection of assembled components for one week yielded the results below, based on the first[2] sample of 40 units from successive lot-by-lot inspection:

Lot No.	Defectives
108	0
109	1
110	2
111	0
112	1
Total	4

---

[2]We are talking of the first sample of 40 from a sequential sample. In general, when sequential sampling is performed, analysis of long-run quality history is best based on the first sample *only*. Subsequent samples are affected by the sampling rules and are thus not truly "random."

Table 2-3.　Master Sampling Table for Continuous Items of Product

LOT SIZE, YARDS	SAMPLE SIZE, YARDS	ACCEPTABLE QUALITY LEVEL																								
		2		3		4		5		6		7		8		10		12		15		17½		20		25	
		A	R	A	R	A	R	A	R	A	R	A	R	A	R	A	R	A	R	A	R	A	R	A	R	A	R
9,999 or Less	525	11	21	16	28	22	36	26	42	32	49	37	55	42	62	53	74	63	86	78	106	92	120	106	135	131	164
	675	15	25	22	34	29	43	36	52	43	60	50	68	57	77	71	92	95	108	104	132	122	150	140	169	173	206
	900	22	32	31	43	41	55	51	67	60	77	70	88	79	99	98	119	117	140	144	172	167	195	191	220	237	270
	1125	29	39	41	53	53	67	65	81	77	94	89	107	102	122	121	142	148	171	183	211	213	241	243	272	301	334
	1275	38	39	53	54	69	70	83	84	95	96	112	113	126	127	154	155	181	182	222	223	257	258	291	292	360	361
10,000 to 24,999	600	12	23	18	31	24	39	30	47	36	54	42	62	48	69	60	83	72	97	90	118	105	135	119	152	150	186
	825	18	29	27	40	36	51	44	61	53	71	61	81	69	90	87	110	104	129	128	156	151	181	170	203	212	248
	975	23	34	33	46	44	59	54	71	64	82	74	94	84	105	105	128	124	149	155	183	180	210	205	238	255	291
	1275	31	42	46	59	59	76	73	90	87	104	100	120	113	134	141	164	167	192	207	235	240	270	272	305	339	375
	1500	43	44	61	62	78	79	95	96	103	104	129	130	144	145	178	179	211	212	258	259	300	301	337	338	418	419
25,000 to 49,999	675	13	25	20	34	26	43	33	51	41	60	47	60	54	76	67	92	81	108	101	131	118	150	135	170	168	208
	900	19	31	29	43	38	55	47	65	57	76	66	87	75	97	93	118	112	139	140	170	163	195	185	220	230	270
	1200	28	40	41	55	53	70	66	84	79	98	91	112	104	126	128	153	153	180	190	220	223	255	252	287	312	352
	1500	36	48	53	67	68	85	85	103	101	120	117	138	133	155	163	188	195	222	241	271	282	314	320	355	396	436
	1800	50	51	72	73	91	92	112	113	132	133	151	152	172	173	211	212	254	255	307	308	356	357	402	403	499	500
50,000 and Up	825	17	30	25	40	34	51	41	60	50	71	58	80	67	90	83	110	100	128	124	157	146	179	165	203	218	249
	1125	25	37	37	52	50	67	60	79	72	93	83	105	95	118	118	145	141	169	175	208	205	238	232	270	289	330
	1425	33	45	48	63	65	82	79	97	94	115	108	130	124	147	153	180	182	210	226	259	263	296	298	336	351	392
	1725	41	54	60	75	81	98	97	116	116	137	133	155	152	175	188	215	224	252	277	310	322	355	365	403	453	494
	2025	56	57	79	80	105	106	126	127	148	149	169	170	192	193	236	237	278	279	345	346	397	398	450	451	556	557

Table 2-4.   Master Single Sampling Table for Continuous
Items of Product

LOT SIZE	SAM-PLE SIZE	ACCEPTABLE QUALITY LEVEL													
		2		**3**		**4**		**5**		**6**		**7**		**8**	
		A	R	A	R	A	R	A	R	A	R	A	R	A	R
9,999 or Less	675	22	23	30	31	38	39	46	47	53	54	62	63	70	71
10,000 to 24,999	825	25	26	36	37	45	46	55	56	65	66	74	75	83	84
25,000 to 49,999	900	27	28	38	39	49	50	59	60	70	71	80	81	90	91
50,000 and Up	975	30	31	41	42	53	54	64	65	74	75	85	86	97	98

LOT SIZE	SAM-PLE SIZE	ACCEPTABLE QUALITY LEVEL											
		10		**12**		**15**		**17½**		**20**		**25**	
		A	R	A	R	A	R	A	R	A	R	A	R
9,999 or Less	675	84	85	100	101	121	122	140	141	158	159	195	196
10,000 to 24,999	825	101	102	120	121	146	147	169	170	191	192	235	236
25,000 to 49,999	900	110	111	129	130	158	159	183	184	207	208	256	257
50,000 and Up	975	119	120	139	140	171	172	197	198	223	224	276	277

Five lots, with a sample of 40 from each, means that 200 units were inspected. A total of four defectives means that the weekly average of defectives, also called *process average*, was 4/200 or 2 percent.

For simplicity, our example utilized a week's data represented by five lots. In practice, a month's inspection results are generally averaged. Comparisons, involving a six months' average and the prior year's average, are also often obtained. Further, performance of several machines or operations, all involving similar equipment and products, may be compared. In instances where process averages are found to represent an unduly high percentage of defective product over the long run, management may have to investigate causes—such as operating methods, age and state of maintenance of equipment, or materials used—and seek remedial action.

In summary then, lot-by-lot inspection provides a continuous surveillance of product quality on a sampling basis. In addition, the quality history built up over a period of time provides management with important information which will often lead to quality, cost, and operating improvements.

REVIEW QUESTIONS

1. Briefly, in what sequence of steps would you proceed in setting up lot-by-lot sampling inspection on a particular product?
2. An assembly line produces 800 small electronic components hourly. One proposal calls for hourly inspection lots of 800, another for half-hourly lots of 400. What important factors should be weighed in making a decision?
3. A shipment of 400 pieces, identified as Part No. 6844, is received. There are four cartons, each containing 100 pieces. Two of them are stamped: Lot No. 10,001 and the other two Lot No. 10,003. A phone call to the supplier informs you that Lot No. 10,002 was shipped to another customer. What would be "rational lots," assuming that (A) you trust the supplier to keep his lots truly separate, (B) you are not sure that a carton contains product from one production lot only?
4. Why do we provide for acceptable quality levels?
5. Contrast single with sequential sampling. What are the major differences, advantages, and disadvantages?
6. An inspector has noted that in the last lot sampled, most defectives were located in the lower right-hand corner. He therefore decides to concentrate his efforts in that corner for the next lot also. Why is

this approach biased and self-defeating? What primary objective of sampling is lost by nonrandom sampling?

7. A gambling casino buys dice in lots of 2,000. An electronic gauge indicates if a die is untrue, which is defined as a failure of any one side to be within a tolerance of 0.001 millimeters. Assuming an AQL of 0.5 percent, what single and sequential sampling plans would be applicable? Next, assuming that your first sample yields (A) zero, (B) one, and (C) two defective (untrue) dice, what action would you take for single and for sequential sampling?

8. Receiving inspection of a subcontracted item involves successive incoming shipments of 2000 parts each, acceptable quality level (AQL) of .75 percent, sample size 225, and acceptance and rejection numbers of 2 and 3 respectively. Given the results below, what is the estimated process average in percent defective?

Lot No.	1	2	3	4	5
No. of Defects in Sample	0	2	4	0	6
Lot Action, Accept/Reject	Acc	Acc	Rej	Acc	Rej

Lot No.	6	7	8	9	10
No. of Defects in Sample	1	1	0	1	2
Lot Action, Accept/Reject	Acc	Acc	Acc	Acc	Acc

Lot No.	11	12	13	14	15
No. of Defects in Sample	0	1	6	2	0
Lot Action, Accept/Reject	Acc	Acc	Rej	Acc	Acc

See also Appendix 3, Case 1, for additional review material.

CHAPTER 3

Basis of Sampling Plans

Although the reader can now apply sampling plans to his quality control problems, he will undoubtedly wish to know something also about the mathematical statistical background on which these plans are based.

Probability and Sampling Plans

We shall illustrate the role of probability mathematics in the building of sampling plans with the aid of an example. Assume that 100 bowls each contain 96 white (or "good") marbles and 4 green (or "bad") marbles. Each bowl is considered to be a production lot or just a lot. The experiment can be readily made by the reader, using just one bowl, but *replacing* the randomly selected beads after each sampling and mixing the bowl well. We wish to find out how it would work out if we use a sampling plan such as the one below:

Sample Size	Acceptance No.	Rejection No.
5	0	1

How many lots will be accepted and how many will be rejected? After 100 samplings, it is most likely that 82 lots will have been accepted and 18 rejected. This may seem odd, since the rejected lots were exactly the same as the accepted ones.

The conclusion may be that the sampling plan is unsatisfactory and that some other plan will give better results. So we may vary the acceptance and rejection numbers and the sample size and start all over again. Soon, however, we will tire of repeated sampling and seek an easier solution.

Calling in the Mathematician

Instead of bothering with sampling trials, the mathematician will approach the problem analytically, using thinking such as follows:

1. Since each bowl contains 4 green and 96 white beads, it is apparent that a sample of just *one* bead will be white 96 times out of 100. The chances, or the probability, of a sample bead's being white will thus be 0.96 or 96 percent.
2. For a sample of *two* beads, the probability of both being white is now 0.96 × 0.96 or 0.92, approximately.
3. Consequently, for *five* beads in the sample, the chances of an all-white sample are approximately:

$$0.96 \times 0.96 \times 0.96 \times 0.96 \times 0.96 = (0.96)^5 = 0.82$$

It follows that about 82 percent of the time there will be a lot acceptance (the sample is all-white) and 18 percent of the time there will be a rejection (there is at least one green bead).

The words "approximately" were applied to the probabilities not just because we rounded out some decimal places. We were aware of the necessity for a small correction in probability values to allow for the fact that *each time a sampling unit is chosen, the composition of the lot is affected* slightly. An exact method for Step 3 would therefore give the probabilities below:

$$96/100 \times 95/99 \times 94/98 \times 93/97 \times 92/96 = 0.81$$

This represents an 81-percent chance of acceptance and a 19-percent probability of rejection of a lot containing 4-percent defectives.

Another Sampling Plan

Among the further questions which the reader may pose for the mathematician, the following is typical: "All right, I accept your calculations. But according to the sampling plan, I would have rejected 19 out of 100 lots that were just as good as the accepted ones. I would like to have a plan with fewer rejections, say just 2 percent. Can you work out such a plan for me?"

"As you wish," is the mathematician's reply. After some figuring, the following plan results:

Sample Size	Acceptance No.	Rejection No.
5	1	2

In practice, this plan will accept lots containing 4 percent defectives 98 percent of the time and reject them 2 percent of the time. The role of probability calculations in building the types of plans desired by the inspector or by management is thus clear.

Translation into Practice

"Now let us go on to some problems that are more like the ones encountered in practice," the reader may say to the mathematician. "Suppose we have a number of bowls, each containing 1,000 beads, but this time we do not know how many beads are green and white. The only thing known is that the lot is to be rejected if it contains more than 4 percent green beads. Is 100-percent inspection necessary in this case, or is there a simpler way?"

"There is," the mathematician replies. "Probability laws can provide us with a sampling plan that indicates with reasonable accuracy whether a lot is acceptable or not. This is such a plan:"

Sample Size	Acceptance No.	Rejection No.
50	3	4

What is meant by "reasonable accuracy"? Obviously, sampling is not so foolproof a way for judging the quality of a lot as is screening. In exchange for the advantage of inspecting only 50 of the 1,000 marbles, we have to take into the bargain a risk of up to 15 percent of rejecting a satisfactory lot. The latter is defined, in our example, as a lot containing defectives corresponding to the acceptable quality level of 4 percent.

"This risk," the reader may respond, "appears too high for practical purposes. I want a sampling plan in which the odds of rejecting a lot of acceptable quality are only 10 percent." Again the mathematician will provide a suitable plan, as given below:

Sample Size	Acceptance No.	Rejection No.
100	6	7

This plan involves larger sample sizes, which means more work for added security. A sampling plan for any desired degree of risk is thus available.

Operating Characteristic Curves

We now come to the final question for the mathematician: "I am going to use your last sampling plan in the shop," the reader says. "There I am confronted daily with lots each containing different proportions of defectives. By using the last given sampling plan, how many acceptances and rejections do I have to expect?"

"That, of course, depends upon the number of defectives in each lot inspected," is the reply. "The operating characteristic (OC) curve of the sampling plan (see Fig. 3-1) will provide the desired answers."

"Suppose your lot happens to contain 10 percent defectives. Refer to the bottom scale labeled *percent defective*, and from the 10-percent point follow the vertical line up, until you reach the curve. On this level, to the right, is the 85-percent point on the vertical scale reading *percent of rejections*. This means that about 85 percent of the time this sampling plan will reject a lot containing 10 percent defectives. On the same level to the left you find the 15-percent point on the vertical scale reading *percent of acceptances*. This means that about 15 percent of the time this sampling plan will accept a lot containing 10 percent defectives.

"Now suppose the quality of the lot happens to be better and it

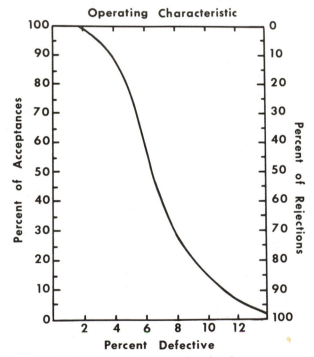

Fig. 3-1. Operating characteristic curve, showing long-run acceptances and rejections that may be expected from use of a given sampling plan. Bottom scale shows possible range of percent of defective product in lots sampled.

contains only 2 percent defectives. Then the curve tells you that this sampling plan will accept such a lot 99 percent of the time and reject it 1 percent of the time."

Every sampling plan may be fully described by its OC curve. For example, in Fig. 3-2 we present four different plans, all based on the same AQL of 4 percent, corresponding to the applicable sequential plans of Table 2-1 of the preceding chapter. A quick comparison reveals that *the steeper the curve, the better is its power to discriminate between good and bad lots.* From this consideration, Curve D is the best one. But can management afford the cost of the large amount of inspection called for by this sampling plan? If not, then a plan such as in C, B, or A may be preferable, despite the progressively poorer OC curve.

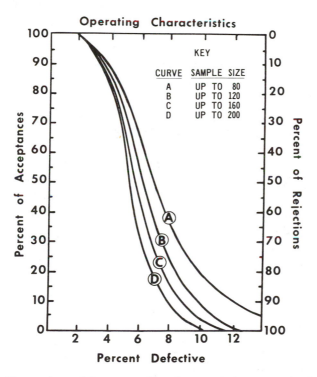

Fig. 3-2. Comparison of four operating characteristic curves. As the sample size increases, a steeper curve results. The steeper the curve, the greater its power to distinguish between acceptable and unacceptable lot quality.

Sampling Risks of Rejection

The single and sequential sampling plans provided in the preceding chapter recognize the following important principles:

1. Samples must be large enough to keep sampling risk low.
2. At the same time, samples must be held within economically feasible dimensions.

The best means of accomplishing this aim is via sequential sampling plans. The plans supplied in Tables 2-1 and 2-2 of Chapter 2 are based on the following risks of rejecting a good lot:

Lot Size	Sequential Sample Size	Approximate Risk of Rejection, Percent
3,199 or less	40 to 200	5
3,200 to 7,999	50 to 300	4
8,000 to 21,999	100 to 500	3
22,000 to 99,999	100 to 1,000	2
100,000 and up	200 to 1,600	1

We define a "good lot" as one having from 1/2 to 1 percentage points fewer defectives than the AQL. The decreasing risks, corresponding to larger lot and sample sizes, reflect the desire to minimize the chances of erroneous rejection of a good lot when the lot is large. An important feature of all sampling, so long as the sample size is no greater than 15 percent of the lot size, governs the risks: It is the absolute sample size that determines the operating characteristic curve. Thus, a sampling plan with a sample size of 50 has the same operating characteristic curve, practically, for a lot of 4,000 as for a lot of 8,000 units.

The probability of rejecting a good lot as a result of sampling fluctuations is also known as the producer's risk. In effect, when a lot of acceptable quality suffers a rejection, the producer (or vendor) incurs losses—a shipment may be returned, needless 100-percent inspection may be required, or other costs and inconveniences may be experienced.

Risks of Acceptance

Corresponding to the chance of erroneously rejecting a good lot, there is also the possibility that a bad lot may be accepted. "Error" here refers to faulty indications of quality of a lot as a result of the

Table 3-1.　Sampling Risks of Acceptance with an Acceptable Quality Level of 4 Percent

Lot Size	Sequential Sample Size	Percent Defectives in Bad Lot				
		5	6	7	8	9
		Percent Risk of Acceptance				
Up to　　499	40 to　　80	78	64	47	35	25
500 to　　799	40 to　120	72	53	36	23	15
800 to　1,299	40 to　160	65	43	26	15	8
1,300 to　3,199	50 to　200	59	34	17	9	4
3,200 to　7,999	50 to　300	48	21	9	3	1
8,000 to 21,999	100 to　500	34	10	2	1	Nil
22,000 and up	100 to 1,600	10	2	Nil	Nil	Nil

"luck of the draw," and does not imply any "mistakes." Since the risk of acceptance varies with what is considered a "bad lot," the data in Table 3-1 have been drawn up to show what happens when a lot with 5, 6, 7, 8, or 9 percent defectives is inspected and the AQL is 4 percent.

The risks of acceptance appear to be large, particularly for the smaller lot sizes. One might well ask how it is possible to maintain an AQL of 4 percent if a lot of size 499 or less incurs 3 chances out of 4 that a lot with 5 percent defectives will be accepted. Moreover, there are 2 chances out of 3 that a lot containing 6 percent defectives will slip by, 1 chance out of 2 that a lot containing 7 percent defectives will go through and 1 chance out of 3 that a lot with 8 percent defectives will be accepted. The answer to this problem is that, so long as there is routine quality control sampling, it is likely that only occasional lots will be of unacceptable quality in the first place. Beyond that, there just is no denying the fact that sampling leaves something to be desired, particularly when sample sizes are not large. Scientific sampling, however, permits us to know the risks incurred beforehand and to choose that sampling plan which is consistent with (1) the economic limitations and (2) the quality protection required. Indeed, if it were not for chance fluctuations of sampling, statistical evaluations and mathematically calculated sampling plans would hardly be needed.

Fortunately, the sampling plans provide a further assurance feature, which will now be discussed.

Average Outgoing Quality Limit

When sampling has rejected a lot, and it is practically and economically feasible to do so, the rejected lot will next be screened to sort out the defectives. When this practice is followed, the (theoretically) perfect lots (screened and sorted) will balance the bad lots that occur occasionally and slip through as a result of the sampling risks just discussed. If the perfect lots are now combined with the imperfect ones, the total percentage of defective product which is finally passed will (for the plans presented in Tables 2-1 and 2-2) not exceed the AQL. In technical language, this feature of the sampling plans is known as average outgoing quality limit (AOQL) protection.

Tables of Sampling Risks of Acceptance

The risks of acceptance shown in Table 3-1 cover plans for an AQL of 4 percent only. In actual practice, another plan, such as for AQL of 0.5 or 2 percent or some other value may be in use. For this purpose, Table 3-2 is provided. Not as detailed as our prior tabulation, with some lesser-used values of AQL omitted, this listing should be adequate in reflecting the important operating characteristics of the sampling plans provided in the prior chapter. For any values not listed in the tabulation, the approximate sampling risks may be evaluated by means of interpolation.

The risks are accurate within ±10 percent. Thus, for example, for an AQL of 5 percent and a lot size 1,300 to 3,199, associated with samples size 50 to 100, we find that a lot containing 8 percent defectives may slip through 20 percent of the time. In actuality, this value may turn out to be 18 or 22 percent.

The problem of sampling and associated risks is thus shown to be a complex one, and statistical methods are indeed essential in setting up a sound sampling program in any organization.

Sampling Risks for Continuous Product

For continuous items of product, Table 3-3 lists the risks of erroneous acceptance of a bad lot. The validity of the sampling risks may be impaired, because (as has been noted) random sampling is

Table 3-2.   Risks of Acceptance Corresponding to Sampling Plans in Tables 2-1 and 2-2

Percent Risk of Acceptance (Percent Defective in Production Lot)

Lot Size	Sample Size	0.5		1.0		2.0		3.0		6.0			8.0			10.0					
		1.0	2.0	2.0	4.0	3.0	5.0	4.0	6.0	6.0	8.0	10.0	7.0	9.0	12.0	10.0	12.0	15.0	12.0	15.0	18.0
499 or less	40 to 80	....	....	60	32	64	32	73	42	76	52	30	81	59	26	75	50	23	72	41	26
500 to 799	40 to 120	....	....	69	28	70	26	58	25	73	35	15	78	43	15	63	30	10	68	30	10
800 to 1,299	40 to 160	45	14	32	10	55	20	55	20	53	26	10	65	38	9	60	19	8	65	22	7
1,300 to 3,199	50 to 200	60	25	34	10	41	10	35	10	54	20	4	60	30	5	45	18	4	55	17	3
3,200 to 7,999	50 to 300	22	6	26	2	37	5	30	4	43	10	1	58	20	2	38	11	2	46	12	Nil
8,000 to 21,999	100 to 500	14	2	17	Nil	28	3	15	2	35	4	Nil	55	10	Nil	21	8	1	43	6	Nil
22,000 to 99,999	100 to 1,000	8	Nil	10	Nil	10	Nil	10	Nil	17	2	Nil	30	3	Nil	....	....	....	....	....	....
100,000 and up	200 to 1,600	2	Nil	3	Nil	5	Nil	5	Nil	10	Nil	Nil	....	....	....	....	....	....	....	....	....

usually not possible on continuous products. The degree of nonrandomness of the distribution of defects will affect the extent to which the risks shown and the sampling procedures provided are applicable. In instances of marked nonrandomness, management may decide that resort should be had to 100-percent inspection as the only acceptable procedure.

Further View of Risk

We may also view sampling risk as the result of an interaction between real world ("true") relationships and sampling observations. The tabulation below will demonstrate:

Management's Action Based on Sampling Observations	Real World Situation	
	The Lot Is "Good"	The Lot Is "Bad"
Accept the Lot	Correct	Type II Error
Reject the Lot	Type I Error	Correct

Type I is the error of rejecting a good lot, Type II is the error of accepting a bad one. Since only one kind of action is possible (accept or reject), we always incur either a Type I or Type II risk. We must also define specifically what percentage of defectives constitutes a "bad" lot, whereas under the O.C. curve approach a range of percentages could be used.

Instead of Type I and Type II, the terms "alpha" and "beta" error are sometimes used. Moreover, in conjunction with vendor-vendee relations, the Type I risk is the "producer's risk," or the risk that the supplier will find his good product erroneously rejected by the sampling "luck of the draw." Type II is the "consumer's risk," representing the possibility that a bad lot may slip through and be erroneously accepted by the buyer (while it should have been rejected and sent back to the vendor or otherwise adjusted).

The concepts just described are useful in giving further insights into sampling relationships. However, in practice, the operating characteristics of a sampling plan give essentially the same information as well as provide data over the entire spectrum of lot qualities that may occur.

When a good lot is accepted, or a bad lot is rejected, a correct decision results. On the other hand, rejection of a good lot or acceptance of a bad lot constitutes one of two types of erroneous action.

Table 3-3.   Risks of Acceptance Corresponding to Sampling Plans in Tables 2-3 and 2-4

Lot Size	Sample Size	ACCEPTABLE QUALITY LEVEL — Percent Defective in Production Lot																							
		2		**3**		**4**		**5**		**6**		**8**		**10**		**12**		**15**		**17½**		**20**		**25**	
Percent Defective in Production Lot		3	4	5	5	5	7	6	8	7	9	9	12	12	14	14	16	17	20	19	22	22	25	27	30
Risk of Acceptance in Percent																									
9,999 or less	525 to 1275	70	20	73	25	85	15	85	18	85	20	88	15	72	21	76	25	75	15	87	25	80	25	85	29
10,000 to 24,999	600 to 1500	65	13	65	20	78	8	82	12	83	15	85	7	65	12	70	17	71	8	86	20	76	15	80	20
25,000 to 49,999	675 to 1800	63	10	62	15	74	5	80	8	81	12	82	5	62	9	63	12	68	5	85	15	75	13	78	18
50,000 or more	825 to 2025	60	8	59	10	70	2	75	5	79	9	79	4	60	8	61	10	62	2	84	12	71	10	73	12

Implications of Sampling

Risk, as we have noted, is an inevitable part of all sampling plans. Moreover, once we have determined the amount of sampling feasible, on the basis of time and cost considerations, the only choice left open is to change our position between the two types of risk. For example, if we wish to lower the risk of rejection, the risk of erroneous acceptance of bad production must automatically rise. Conversely, a decrease in faulty acceptances can be accomplished only by increasing the risk of erroneous rejection of good product. Each of the two types of risk has major cost implications. Rejection of a good lot entails these problems:

1. The flow of product and the operation of equipment are impeded or stopped until a decision is reached.
2. Good equipment and machinery may be unneccessarily subjected to costly checks, replacement of parts, or various maintenance work.
3. Lots that are still within acceptable quality levels are needlessly subjected to 100-percent inspection.

Aside from the immediate cost in dollars and cents, there are the more difficult-to-measure losses in worker morale or the confidence of the foreman in the inspection and quality control operations. Accordingly, we wish to make sure that erroneous rejections occur as seldom as possible.

Now let us consider the drawbacks of erroneous acceptance of an unsatisfactory lot:

1. Bad product is permitted to continue through further production. It may be assembled into complex components and produce malfunction of the final product.
2. Sometimes, unsatisfactory product may not be discovered at the plant, but will eventually give rise to customer complaints or the request for field visits and repairs.
3. Instead of off-standard product's being returned to a vendor, it is accepted and paid for by the plant, causing production problems at a later stage.

Here again we incur monetary as well as intangible losses, and it is difficult to say which type of error—rejection of a good lot or acceptance of a bad one—can cause more problems.

One may easily become depressed from a lengthy consideration of all of these factors. Such an attitude is not, however, warranted. In general, we do *not* install sampling inspection as a corrective device; screening or 100-percent inspection performs that function. Instead, lot-by-lot inspection applies to lots that come from a process that does normally conform to an acceptable quality level (AQL). Consequently, the occurrence of off-standard lots is to be expected rarely. The choice of the risk of acceptance of bad product, on the other hand, has been placed at relatively low levels. Erroneous rejections, therefore, are likely to be at a low enough frequency to hold expenses and intangible costs at a relative minimum.

Full Set of Operating Characteristics

Sampling risks of acceptance of a bad lot or rejection of a good lot have been tabulated on pages 36 and 39, providing the key risk values of interest. On occasion, however, it is desirable to have all of the risk values available for review of the practical range of lot qualities. Accordingly, Appendix 2 provides all of the Operating Characteristic (OC) curves pertaining to the sampling plans in the master Tables 2-1 and 2-2.

REVIEW QUESTIONS

1. By drawing random samples from a bowl, we can simulate the effects of lot-by-lot sampling in production. Of what value is it to examine the actual or likely outcomes from such bowl experiments?
2. A bowl contains 900 green and 100 red beads. What is the probability of drawing (A) a green bead or (B) a red bead in a random sampling of one bead?
3. In sampling, what is (A) the sampling risk of rejection and (B) the sampling risk of acceptance?
4. How can one reduce both—the risk of rejection and acceptance—in sampling?
5. What is the value of the operating characteristic curve of a sampling plan?
6. In what manner are tables of the sampling risk of acceptance a useful substitute for the operating characteristic curve?
7. Is the concept of average outgoing quality limit (AOQL) of value in destructive testing?
8. A plastic cover plate must withstand an impact of 5 kilograms dropped from a height of 1 meter. Assuming an AQL of 4 percent, a sample size (sequential) of up to 80, and 6 percent defectives in the lot, what is the probability of (A) accepting or (B) rejecting the lot?

9. For the results of the lot-by-lot inspection of the deburring operation
   of the preceding chapter (Review Question 8, page 30), an alternate
   means of estimating the process average is by using the OC curve
   applicable for the particular sampling plan. Using the OC curves in
   Appendix 2, what is the estimate of the process average, and why is
   it different from the earlier estimate?

CHAPTER 4

Process Inspection with Control Charts

When processes and operations require close surveillance of quality during production, process inspection is best accomplished with the aid of control charts.

Purpose of Control Charts

The primary purpose of control charts is to show trends in weight, dimensional, electrical, or other characteristics of product toward whatever maximum and minimum limits mark the dividing lines between acceptable and unacceptable quality levels. While the control chart also indicates when established limits have been exceeded, its main purpose is to provide the means of anticipating and correcting whatever causes may be responsible for defective products. Prevention of defective output, rather than mere discovery and correction of unsatisfactory work, is the fundamental principle involved.

Graphic Records of Tests

The working of a control chart will be presented by way of an example. In Fig. 4-1 we begin with a simple graphic record of hardness tests on successive outputs of shovel blades. Close control is needed, since a soft blade will bend and a hard blade will be so brittle that it will break in use.

From experience, a hardness of 45 $\pm$ 3 has been specified. This means that the desired hardness for each blade is 45, but that product measuring anywhere between 42 and 48 is still within the specification limits and hence conforms to the tolerances. From each successive heat, a sample of four blades is selected randomly, and each test result is plotted on the graphic record. For the first sample, therefore, we plotted the readings 46, 45, 45, 44.

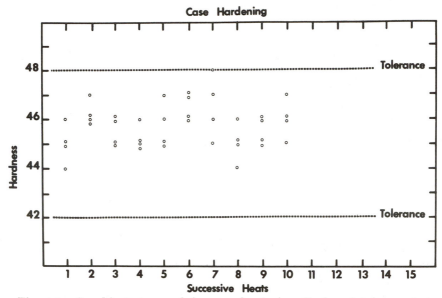

Fig. 4-1. Graphic test record for case hardening. Each point is one test result. Four workpieces are tested per heat. Upper and lower tolerance limits are indicated.

The variation in the sample reflects the variation in the production process. Typical causes of such variation are as follows: When the blades are in the furnace, they may not all be exposed to exactly the same amount of heat; they may not all be kept in the furnace exactly the same length of time; and there is undoubtedly some variation in the metal itself. In addition, there are differences in pressure used in testing each blade.

If we were not dealing with a furnace, but with a precision machine, we would still have tool wear, play in bearings and cams, maladjustments in settings, vibration, and atmospheric conditions which would cause variation. Or if we were winding armatures, there would be variation in winding tensions, wire thickness, and coating uniformity. The very fact that a production process requires quality surveillance means that noticeable and potentially troublesome causes of variation are present.

Plotting Sample Averages

In order to simplify the readings of the test values, it is customary to plot just one point for each sample, such as our sample of four blades per furnace heat. This point is called the *sample average*. For the first sample, we obtain:

Test No.	Test Result
1	46
2	45
3	45
4	44

Total	180
Average, 180/4	45

When the averages are plotted, the new chart will appear as portrayed in Fig. 4-2.

Control Chart

When control limits are inserted on the graphic record of sample averages, a control chart results, as shown in Fig. 4-3. It will be noted that the control lines form a narrower band than the tolerances derived from the specification. If we were to demand that all

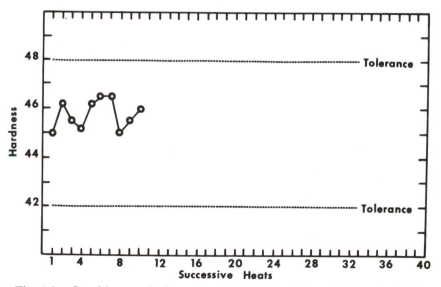

Fig. 4-2. Graphic record of averages for case hardening. Each point is the average of a sample of four workpieces tested.

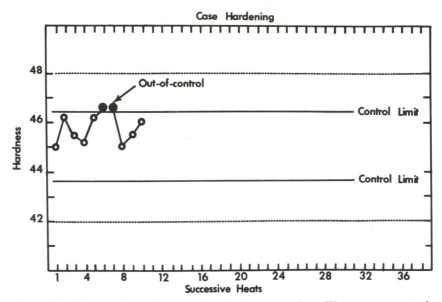

Fig. 4-3. Control chart for case-hardening operation. The upper control limit and lower control limit are indicated. Two out-of-control occurences are noted.

product should fall within this narrow band, this would be equivalent to setting more exacting demands on the process than indicated by the production specification. This is not the intention. All the product within tolerance limits still is acceptable. Certain conclusions are drawn, however, if a sample average should fall outside the control lines.

What the Control Chart Shows

Examining our control chart, we note that most of the time the plotted sample averages are well within the control limits. In two instances, the plotted points fell outside the control limits. Such an out-of-control point has only one significance: something has probably gone wrong with the process, and a check is in order to find and correct the cause. The fault may be simple. It may be a maladjustment of the furnace heat, improper timing, careless placement of the blades, or the use of inferior metal.

Once a cause has been eliminated, the subsequent plotted points will stay well within limits. But if more "outsiders" appear, then a very thorough investigation should be made, even if this means temporarily shutting down operations until proper adjustments and corrections have been effected.

The reader may ask why it took *two* plotted points for management to check and correct the cause of unsatisfactory output. The answer to this is that the ten plotted points on the chart represent *past data* that were used to set up the chart and the limits. But the control chart is now ready for hourly use *on a current basis*. Consequently, in the future, the occurrence of just *one* out-of-control point will signal the urgent need for immediate corrective action. Moreover, by watching the *trend of successive points*, we may spot *impending trouble* before an off-limits condition arises.

Further Applications

The wide diversity of applications of control charts in production will be illustrated by a few further examples.

Control of the diameter of a machined part is illustrated in Fig. 4-4, which is typical of the problems experienced on certain turret lathes, grinding operations, and other processes. In some instances, operators tend to change tools too often or to make unnecessary

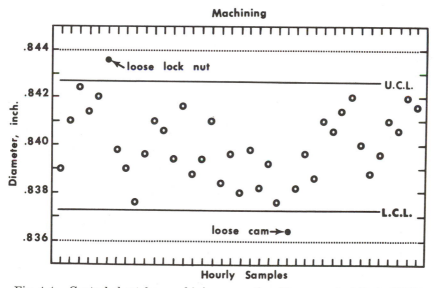

Fig. 4-4.   Control chart for machining operation. Upper control limit (UCL) and lower control limit (LCL) are shown.

adjustments. Others tend to run the machine too long without sharpening tools or they operate equipment carelessly. The remedy is to have a control chart right at the machine, which tells the operator:

1. So long as plotted points are within control, no corrections are necessary.
2. When a series of points shows a trend toward an upper or lower limit, get ready to make corrective adjustments before out-of-control product appears.
3. If, despite careful work, out-of-control points appear, and correction is not a matter of simple machine adjustments or tool changes, report the machine for repair and maintenance work.

Additional examples of quality control charts appear in Figs. 4-5 and 4-6. In all instances, the purposes are the same: to maintain a surveillance of the process and to spot and correct impending trouble or off-standard production as soon as possible. Because control charts have demonstrated their power in aiding management, supervision, and the operator in controlling production processes, their value is widely recognized.

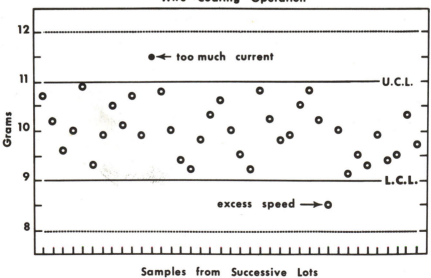

Fig. 4-5.   Control chart for wire-coating operation.

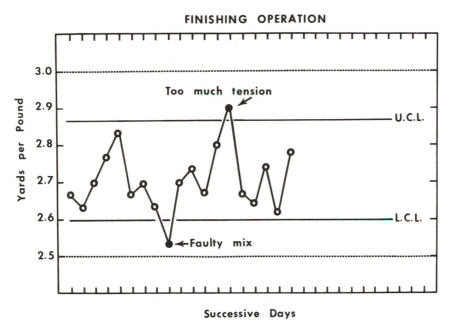

Fig. 4-6. Control chart for finishing operation. Weight in yards per pound of finished fabric is recorded. Faulty proportioning of chemical admixtures and excessive finishing roll tensions caused the out-of-control points shown.

Process Variability

You cannot locate control lines without knowing about the variability of the process. The concept of variability is derived from the following considerations:

No process can turn out identical product from unit-to-unit, hour-to-hour or day-to-day. Even with the best equipment and the most skilled operation, no two units of product will be exactly alike. It is for this reason, for example, that hardnesses of shovel blades vary, that machined parts differ, and that variation is observed in other operations. The variation observed in the product reflects the variability of the process.

While one cannot eliminate this natural variability, one can try to keep it within the requirements of the specification limits, assuming that the latter have been set realistically.

Measurement of Variability

There are many ways to measure process variability. One of the simplest and generally most practical approaches is to use the average of a series of sample ranges. This method will be illustrated with reference to the prior example of hardness testing. Recall that the first sample yielded four values: 46, 45, 45, 44 on a Rockwell C-scale. The difference between the highest value, 46, and the lowest, 44, is the sample range R of 2.

Repeat the determination of R for the next nine samples (or more). The data are recorded in Table 4-1. The sum of the ranges is 16. Dividing this total by the number of ranges, 10, yields the *average range*, $\overline{R}$, of 1.6 as a measure of process variability.

The importance of analyzing and evaluating process variability is further underscored in Fig. 4-7, demonstrating how the quality of a lot is directly affected by the spread of the test results. Although both lots have the same average, the lot exhibiting greater uniformity (less variability) is the better lot.

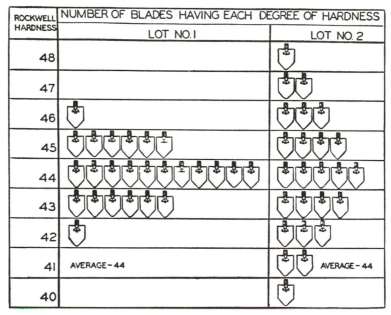

Fig. 4-7. Comparison of variability. Lot No. 1, exhibiting greater uniformity, is the better lot. As a result, although both lots have the same average, only Lot No. 2 (with the higher spread) has product below the lower tolerance of 42.

Table 4-1.   Analysis of Process Variability
(Quality Characteristic: Rockwell Hardness, C-Scale)

Sample No.	First Blade	Second Blade	Third Blade	Fourth Blade	Sample Range	Sample Average
		Test Results				
1	46	45	45	44	2	45.00
2	46	47	46	46	1	46.25
3	45	45	46	46	1	45.50
4	46	45	45	45	1	45.25
5	46	47	47	45	2	46.25
6	46	46	47	47	1	46.50
7	46	48	47	45	3	46.50
8	45	46	44	45	2	45.00
9	45	46	46	45	1	45.50
10	46	47	46	45	2	46.00
Totals					16	457.75

$$\text{Average Range, } \overline{R} = \frac{\text{Total of Ranges}}{\text{No. of Samples}} = 16/10 = 1.6$$

$$\text{Process Average, } \overline{\overline{X}} = \frac{\text{Total of Sample Averages}}{\text{No. of Samples}} = 457.75/10 = 45.8$$

Process Average

The entries in Table 4-1 also yield sample averages, $\overline{X}$, which can be totaled. The sum so obtained is 457.75. Since there were ten samples, we again divide by 10, this time obtaining a grand average (average of the averages) of 45.8 (rounded). The symbol for this is an X with two bars over it, $\overline{\overline{X}}$, which may be called the *process average*. It should be noted, however, that both the average range and the process average are estimates. Since we have sampled, we do not know the true process average. All we can do is hope that the sample values give a reasonably good estimate. This assumption is usually justified.

The process average is, in effect, an average of individual sample averages. For this reason, the terms *grand average* and *grand average of the process* are often applied to this value.

Calculation of Control Limits

Having calculated the average range, we are ready to begin the development of control limits. The procedure is quite simple, as the four steps below demonstrate:

Step 1: From a series of sample ranges, calculate the average range $\overline{R}$ as a measure of the variability of the process.

Step 2: Multiply the average range by a factor, depending upon sample size, and provided in Table 4-2. Since we tested four blades per sample, the sample size is 4 and the control-limit factor is 1.02 or 1.0 (rounded). We obtain:

$$\overline{R} \times F_s = 1.6 \times 1.0 = 1.6$$

The term F_s is an abbreviated form for "Control Limit Factor."

Step 3: Subtract the value found in Step 2 from the upper tolerance limit and add it to the lower tolerance limit. The result obtained represents the upper and lower control limits (UCL and LCL). We have:

$$\text{UCL} = \text{Upper Tolerance} - 1.6 = 48 - 1.6 = 46.4$$
$$\text{LCL} = \text{Lower Tolerance} + 1.6 = 42 + 1.6 = 43.6$$

Step 4: Draw the control limits of 46.4 and 43.6 on the control chart. In our illustrative example, the first ten samples represent the past data. They are of little interest now. The control limits are extended beyond the past data, for the purpose of using the chart to plot averages from successive future testing. The control limits then provide ready criteria for the evaluation of quality of current production.

The task of developing a control chart has thus been illustrated with the aid of an example.

Practical Hints

Despite the multitude of successful control chart applications, problems and even outright failures may arise, particularly when formulas are used without a consideration of the practical limitations involved. One must be alert, therefore, to the following types of questions and problems that may come up:

1. Have the specification limits or tolerances been set up realistically? Statistical methods are available for the analysis of specifications (Chapters 8, 14, and 15), and they should be

Table 4-2.   Control Limit Factors for Charts for Averages
Based on Specification Limits

Sample Size	Factor, F_s	Rounded Value of F_s
2	1.49	1.5
3	1.14	1.1
4	1.02	1.0
5	0.94	0.9
6	0.90	0.9
8	0.84	0.8
10	0.80	0.8
12	0.77	0.8
15	0.74	0.7
20	0.71	0.7

NOTES:
1. Sample size represents the number of observations (test results) per each sample. Thus, if a sample consists of four components or other parts tested, the sample size is 4.
2. For many practical purposes, particularly when no calculator is on hand, the rounded values of F_s are often considered adequate.
3. The term F_s means *control limit factor*, and the subscript s denotes that the control limits are computed with reference to specification (or tolerance) limits.
4. A derivation of these factors is given in Chapter 6.

used in all instances where doubts about current limits exist. Realistic tolerances are based on requirements that consider the desired and necessary performance characteristics of products, and give due regard to the capability of the process to meet the tolerance requirements.

2. Is the process variability estimated from sound samples? From 10 to 50 samples, from the immediate past, should be used. While larger numbers of samples provide more data and thus better estimates, it should also be remembered that taking more samples means going back farther into old test results. The older values may no longer be representative of current process performance capability. A good compromise between "lots of data" and "very recent data" needs is to utilize from 20 to 25 samples. Samples should, of course, be based on random selection. Moreover, any markedly off-standard production should not be included in the calculations.

3. Are the samples rational? A rational sample can be readily identified with a particular time and source, so that tracing of causes of unsatisfactory output is facilitated.

4. Are the control limits up-to-date? Once established on a par-

ticular production operation, control limits are generally useful for several weeks or months. At regular intervals, however, there should be a recomputation of the average range to note whether or not the process has undergone any changes in the course of time.

The last point is particularly interesting. With the use of control charts, there is often a gradual but significant improvement in the variability of the process. A reduced process spread may often mean that the firm can now bid for close-tolerance work, where previously this would not have been feasible. In other instances, a narrowed process variability will permit savings in costly raw materials or increased productivity.

Multidimensional Operations

There are many processes where more than one quality characteristic may require control, such as multidimensional operations on a turret lathe. In many instances, management finds it impractical to install and use a multitude of charts on just one machine. In such cases, it may be preferable to confine oneself to the one or two most important characteristics. The remaining dimensions are subjected to occasional spot checks.

The drive toward simplicity may go too far at times. For example, in the production of an elastomer, strength was the most important characteristic. Control chart use, therefore, was confined to this quality test. Only after a good many rejections by customers was it realized that elongation should also have been charted. In fact, it was discovered that elongation will often exhibit a reverse trend from tensile values. In other words, high strength (a desirable characteristic) can result at the expense of lowered elongation characteristics (an undesirable aspect).

Sampling Last Work Produced

Lot-by-lot inspection, it will be recalled, requires that samples be based on a random selection from all parts of a particular lot. Under process inspection, however, which is primarily concerned with successive or continuous runs of production (rather than separate and distinct lots), the rule of random sampling need not apply rigidly. Instead, it may be equally feasible or even preferable to take workpieces directly from the machine. For example, the inspector may check a sample of four units produced just prior to his arrival at a machine. This method has the advantage of providing the *latest information* about quality, which is of great importance in detecting any impending trouble.

It is essential, of course, to use only the last regular production and to guard against samples "stacked" in advance with only the best units. Never should the sample represent work produced in the inspector's presence—unless automated production is checked.

In the calculation of control limits, the average range should *always* be based on random samples, so as to fully reflect the variability of the production process. The potential chain of errors is as follows:

1. Last-work samples are likely to show less spread than random samples.
2. Estimates of process variability will now be understated.
3. Control limit bands, based on understated average range, will be wider than they should be.
4. Greater width of the limits means lowered sensitivity of the control chart.
5. Lowered sensitivity, in turn, increases the risk of erroneously accepting off-standard quality of production.

Consequently, while last-work samples may be quite beneficial for current control of production; control chart limits should be based on variability ascertained from (additional) random samples.

A Technical Note

The type of control chart just presented is only the first of several that will be covered in this book. It has been selected because of its practical usefulness. It is known as a "control chart for averages" because we plot sample averages. Since there are several kinds of control charts for averages, we should amend the label to say "control chart for averages with specification-oriented limits." The term "modified control limits" is also found in the literature.

REVIEW QUESTIONS

1. In what manner can a control chart help in the prevention of nonconforming product?
2. Why do we prefer averages to individual units when preparing control charts based on small samples?
3. When a control chart for sample averages has been prepared and is plotted on a current basis, what information will it reveal?

4. Compare process average and process variability. How do they differ? Why do we need to know the magnitude of both?
5. Although we stressed random sampling in lot-by-lot inspection, in process inspection with the aid of control charts it is often preferred to confine sampling to the last few units produced. How can such non-random sampling be defended? What precautions must be taken?
6. A nonrandom sample may still be a rational sample. How is this possible?
7. It is said that the average range reflects process variability. Why is this true?
8. The overall dimension of a component must be 60 millimeters with a tolerance of ±0.08 millimeters. The test results below were found:

June 12	+0.02	+0.01	0.0
June 13	−0.01	+0.03	+0.01
June 14	−0.03	+0.01	+0.04
June 15	+0.03	−0.06	−0.03
June 16	+0.01	0.0	−0.01

What are the (A) process average and (B) average range, based on these five samples of three pieces each? Next, what are the upper and lower control limits?

See also Appendix 3, Case 2 for additional practice with control charts. Assume, for the purpose of this case, that Specification Limits are at 3300 ± 40 ohm.

Control charts may also be developed for Case 3, assuming for this purpose a tolerance of 34.5 ± 1.0 seconds.

CHAPTER 5

Control of Variability

Many situations arise in which it is desirable to supplement the recording and graphing of sample averages with accompanying plotting of sample ranges. In this way, an hourly, daily, and weekly surveillance over process variability is afforded. The control chart for ranges serves this purpose.

Applicability

A range control chart should always be used when process variability is unstable and unpredictable—that is, when sudden increases in variability may occur at any time. An obvious area of need represents applications on old equipment, where machine parts are worn or tend to go out of alignment frequently. Even new equipment may be quite variable, when extreme demands for quality and uniformity of output are pushed toward the very limits of a current state of technology. The range chart does not replace the chart for sample averages; it merely supplements the latter with additional information.

To illustrate, let us assume that variability on a screw machine operation has suddenly increased because of a loose cam. While the average dimension of the product may still be acceptable, the *individual parts* vary so greatly that some of them begin to fall outside the tolerances. The control chart for averages will not yet indicate trouble, but the range chart will, because it is specifically designed to detect undue variability.

Preparation of Range Chart

The distinguishing feature of the range chart is the control limit for ranges. If a sample range exceeds the control limit, it thereby signals the presence of excessive process variability. In Fig. 5-1, we present our earlier case-hardening illustration, with a range chart

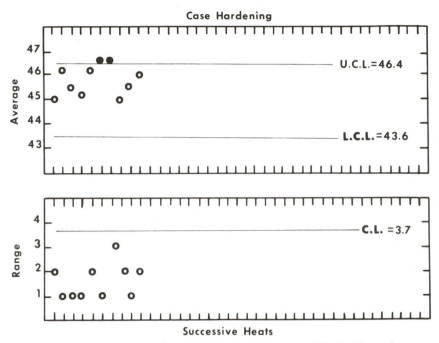

Fig. 5-1.   Chart for control of averages and ranges. Illustration of case-hardening operation.

added. It will be observed that variability stayed in control. The calculation of the control limit is simple:

1. Note the average range $\overline{R}$, obtained from calculations for the chart for sample averages. In our example $\overline{R}$ is 1.6.
2. From Table 5-1, obtain the control limit factor. For samples of size 4, the control limit factor, D_4, is 2.3.
3. Obtain the control limit by multiplying $\overline{R} \times D_4$. Hence, the control limit is 1.6 $\times$ 2.3 or 3.7.

Insert the limit on the chart and use it in daily control of the uniformity of production.

The limit found is an upper one. When sample sizes exceed 6, it is also possible to obtain a lower limit. For this purpose, substitute factor D_3 for D_4 in the calculations. Can variability ever be too low ? Not really, because we prefer uniformity. But an unduly low sample range may indicate imprecise testing or recording of instrument

Table 5-1. Control Limit Factors for Range Charts

Sample Size	Factor D_4 for Upper Limit		Factor D_3 for Lower Limit	
	Detailed	Rounded	Detailed	Rounded
2	3.267	3.3	. . .	. . .
3	2.575	2.6	. . .	. . .
4	2.282	2.3	. . .	. . .
5	2.115	2.1	. . .	. . .
6	2.004	2.0	. . .	. . .
8	1.864	1.9	0.136	0.14
10	1.777	1.8	0.223	0.22
12	1.716	1.7	0.284	0.28
15	1.652	1.7	0.348	0.35
20	1.586	1.6	0.414	0.41
25	1.541	1.5	0.459	0.46

SOURCE: Factors D_4 and D_3 are provided in *Manual of Quality Control of Materials,* American Society for Testing and Materials (Philadelphia, 1951), p. 115. The usually justified assumption of reasonably normal distribution of lot quality is made. The topic of normal curve will be treated further in the next chapter.

NOTES ". . ." means "no factor applicable, control limit is zero." Sample size represents the number of observations (number of units tested or measured) in each sample. For example, if four separate snap links are tested for strength, resulting in four strength values, the range is based on a sample size of 4 units of product. In other words, the sample size is 4.

readings. Also, if a process by chance shows continued low variability, we may like to be alerted to this fact, so that we can find the reason and try to adjust operating procedures toward this desirable result.

Further Illustration

In Fig. 5-2, an example is given of a range chart for the control of variability in a centerless grinding operation. When the first out-of-control range occurred, the output produced was subjected to 100-percent inspection, which revealed a sizable proportion of defectives. Yet, a check of the equipment could not reveal anything wrong. Management decided to increase the wheel dressing frequency. A short while later, however, another out-of-control situation developed. A renewed check of equipment still did not locate any trouble spot. Thereupon an investigation was made of the diameter of the work being fed into the machine. This revealed that it did not generally come in a very satisfactory condition from the preceding turning operation. Sporadic occurrence of excessive variability of the turned parts was thus established as the cause of trouble.

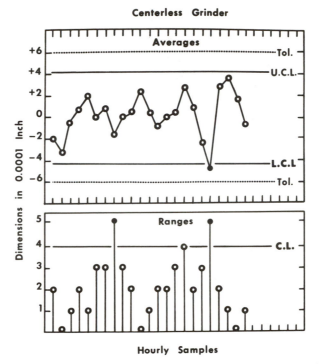

Fig. 5-2.   Control of average level and variability on centerless grinder.

The remedy was to install control charts on the turning operation. As a consequence, the grinding operation was likewise brought into control. In time, moreover, a definite decrease of variability in the grinding operation was also accomplished. The more uniform incoming product permitted better operation of the grinder.

Note also, from the accompanying chart for sample averages, that (1) average levels may be satisfactory while variability goes out of control and that (2) both average level and variability may go out of control simultaneously. The latter usually means worse problems.

Further Calculations

Let us now demonstrate the calculations leading to the control limits for the centerless grinding operation. We have the following given data, all in 0.0001 inch:

Tolerance Limits: $+6$ and -6.

Nominal Dimension: 0

Average range, $\overline{R}$, from recent past experience: 1.9

Sample size: 5

Tabulated factor D_4 (for sample size 5): 2.1

Tabulated factor F_s (for sample size 5): 0.9

We proceed as follows:

1. Control Limit for Ranges

   Multiply $\overline{R} \times D_4 = 1.9 \times 2.1 = 4.0$
2. Control Limit for Averages

   Multiply $\overline{R} \times F_s = 1.9 \times 0.9 = 1.7$

   Upper Control Limit $=$ Upper Tolerance $- 1.7$
   $$= 6 - 1.7 = 4.3$$

   Lower Control Limit $=$ Lower Tolerance $+ 1.7$
   $$= -6 + 1.7 = -4.3$$

These are the limits portrayed on the control chart for averages. We note from the tabulated values of D_3 that for a sample size of 6 no lower control for the range chart applies. Factors are given for larger sample sizes, and in such instances, we would have merely substituted D_3 for D_4 to obtain lower control limits for ranges.

Patterns of Variability

The effect of process variation may be further elucidated with the aid of frequency distributions, such as that portrayed in Fig. 5-3 for a coil-winding operation. The base scale shows the quality measurement, which is resistance in ohms as shown (plus 1,000 ohms). For example, the center bar of the lower part of the diagram shows the number of coils that measured 50 + 1,000 or 1,050 ohms. To either side, there are columns showing the results for 1,045 and 1,055 ohms.

The height of each column represents the frequency of occurrence of each measurement. Again for the lower section of the chart, the value 1,050 was observed 70 times. The value 1,055 occurred 60 times, while 1,045 was measured 55 times. The total of all frequencies is 300, which means that 300 coils were tested. The product of the lower section is well centered within tolerances. The 300 coils of the upper section, however, present a different picture. The product is still centered around 1,050 ohms, but variability in processing has produced an undue spread of measurements. Off-standard product, exceeding both tolerance limits, results.

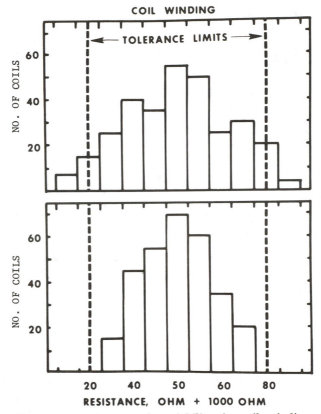

Fig. 5-3. Frequency patterns and variability in coil winding. In lower section, pattern is well within control. In upper section, distribution has broadened, resulting in out-of-tolerance product. Purpose of range control chart is to identify increases in variability when they occur.

The reader may readily imagine a third condition, in which product exhibits the variability pattern of the distribution in the lower section, but the entire distribution moves toward the higher side of the scale. All columns, including the central column, will be shifted in that direction. Off-standard product now results as soon as the upper part of the distribution crosses the upper tolerance limit.

When we use control charts, we avoid the labor of testing large amounts of product and instead rely on a relatively small sample. From this sample, a judgment is made regarding the output as a whole. The following parallels apply:

1. A plotted sample average and sample range both fall inside control limits. The likely distribution pattern is similar to a situation where product is well centered and well within tolerances (lower section of the diagram).
2. A plotted sample average is within control, but the sample range is out-of-control. The likely distribution is properly centered but has spread excessively, thereby producing output beyond the tolerance limits (upper section of diagram).
3. A plotted sample average is out-of-control, but the sample range is in control. The likely distribution has remained relatively uniform, but it has shifted to either the right or left side of the base scale, thereby producing defectives (proportion of product outside either one of the tolerance limits).
4. Both sample average and range are out-of-control. This is the worst type of situation. The average level of the process has shifted, and at the same time the process has broadened excessively. Large quantities of off-standard product result.

Control Policy

As we have noted, after a range chart has been maintained for some time, it may lead to a location and elimination of major sources of excessive variability. In the example just presented, lack of proper uniformity of incoming material was identified as the cause and eliminated. At other times, the range chart will supply evidence that a more regular schedule of equipment maintenance should be adopted. On occasion, it may be found that certain processing sequences should be altered or that some machine components need replacing.

Prior to the plotting of a range chart, these needs are often not recognized clearly enough to permit the making of strong recommendations or the attainment of prompt and vigorous corrective action.

Once the primary sources of variability have been found and eliminated, one may well consider eliminating the routine use of a control chart for ranges (though still maintaining a chart for sample averages). In fact, in a good many organizations there is a policy to continually review (1) the need for new charts and (2) the elimination of charts that have become outdated. Unnecessary charts should not be kept in use. They represent needless expenditure of time and effort, and they tend to detract interest from the really useful and active charts.

Related Charts

The range is not the only measure of variability. Another, more accurate value is a figure known as "standard deviation" and denoted by the symbol σ (a small Greek sigma). We will demonstrate its calculation in the next chapter. From this work it will be recognized that a price is paid for greater precision of estimation: the calculations become more lengthy.

For these reasons, although the evaluation of σ becomes important for many quality control purposes, it is usually not a useful and practical measure in ordinary control chart applications. There are some specialized situations in which control charts for sample standard deviations are of value, but we shall omit this relatively rarely used approach.

Concluding Observation

Range charts are of considerable help in quality control work whenever routine surveillance over process variability is desired. The range chart is usually used as a supplement to the control chart for averages. However, when process variability is relatively stable and predictable, there is little justification for maintaining a range control chart.

REVIEW QUESTIONS

1. Under what conditions would you supplement a control chart for sample averages with a range control chart?
2. What factors would you check if you encountered an unduly low variability (point below the lower control limit on a range control chart)?
3. What conclusions would you draw if (A) the sample average is in control *and* (B) the sample range is out-of-control.
4. Frequency distributions reveal the pattern of variability of a production process. Why, then, should we use also the range control chart?
5. An assembly line produces 50 units per hour. Current quality control policy is to maintain and plot control charts for averages and ranges on a half-hourly basis. In addition, the 400 units accumulated at the end of the shift are subjected to lot-by-lot sampling. Before a lot may move to the next department, it must be found acceptable on the basis of the control chart record and the lot-by-lot sampling results. Unacceptable product is subjected to 100-percent screening inspection.

Do you consider this approach to be overly cautious? (Discuss the conditions under which *you* would or would not recommend such an approach.)

6. Where would you keep control charts, in the inspection department or on the production floor? Why?

7. For the dimensional values of question 8, prior chapter, what are the upper- and lower-range control-chart limits?

See also Appendix 3, Cases 2 and 3, for which control charts for variability are applicable.

CHAPTER 6

Statistical Basis of Control Charts

The theoretical foundations of the practical control charting methods will now be discussed. We begin with the concept of *standard deviation*.

Standard Deviation

The standard deviation serves as an indicator of the variation exhibited by a frequency distribution, such as shown in Fig. 6-1 for the burst strength of 40 lengths of plastic tubing which was tested with the results shown.

From lines *a* and *b* it is noted that one tube tested 22 pounds per square inch, psi, in terms of burst strength. Three tubes tested 23 pounds, and 9 tubes tested 24 pounds. Next, 14 of the tubes tested 25 pounds, 9 tubes tested 26 pounds, and so on. Although in this illustration a burst strength of 25 pounds is readily recognizable as the distribution average, usually it will have to be computed (for less symmetrical patterns of distributions). For this purpose, line *c* is used, which is obtained from successive multiplications of $a \times b$. For example, $22 \times 1 = 22$, $23 \times 3 = 69$, and $24 \times 9 = 216$. The total of this "frequency $\times$ strength" line is 1,000 psi. Next, division by the total number of units tested, 40, gives $1,000/40 = 25$ as the average or arithmetic mean in terms of psi.

In line *d*, the deviations of each value in *a* from the mean are noted. Thus, $22 - 25 = -3$, $23 - 25 = -2$ and $24 - 25 = -1$. The next step, now, is to eliminate minus signs by squaring each deviation. For example, -3 squared is $+9$. Finally, the squared values in line *e* are again multiplied by the frequency in *b*. The total of line *f*, containing the results of these multiplications, is 60 psi. Again, in obtaining an average—this time of the squared deviations—we find:

$$\text{Variance} = 60 \text{ psi}/40 = 1.5 \text{ psi}$$

69

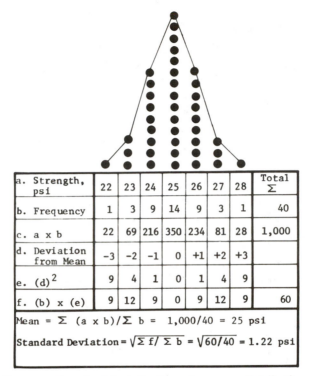

a. Strength, psi	22	23	24	25	26	27	28	Total Σ
b. Frequency	1	3	9	14	9	3	1	40
c. a x b	22	69	216	350	234	81	28	1,000
d. Deviation from Mean	-3	-2	-1	0	+1	+2	+3	
e. (d)^2	9	4	1	0	1	4	9	
f. (b) x (e)	9	12	9	0	9	12	9	60

Mean = Σ (a x b)/Σ b = 1,000/40 = 25 psi

Standard Deviation = $\sqrt{\Sigma f / \Sigma b}$ = $\sqrt{60/40}$ = 1.22 psi

Fig. 6-1. Standard deviation for a frequency distribution.

Since the deviations had been squared, it is now necessary to obtain the square root. Hence,

$$\text{Standard Deviation} = \sqrt{\text{Variance}}$$
$$= \sqrt{1.5} \qquad = 1.22$$

Observe that the term *Variance* refers to the squared standard deviation. The latter is commonly denoted by σ, a small Greek sigma. We may look on the variance as a stepping stone in the calculation of σ, but in certain more advanced applications the variance is used in its own right (see Chapter 12).

Value of Standard Deviation

The standard deviation is a numerical measure reflecting in a single number the nature of variation in a set of data. The larger the σ, the broader will be the frequency distribution pattern. For example, Fig. 6-2 shows a pattern with the same average as the one just examined, but it is a more dispersed distribution with a wider spread. The greater σ of 1.70 psi (as against σ of 1.22 just above) reflects this fact.

Normal Curve

Experience in industrial production confirms that a phenomenon, widely encountered in the sciences and all aspects of life, also holds for the frequency distribution pattern of products produced: quality characteristics tend to conform to a so-called *normal curve*. Figure 6-3 illustrates this bell-shaped frequency pattern. Figure 6-4 shows a few examples of industrial products and their distribution, demonstrating how these generally conform to normality. Such a (relatively small) lack of agreement between an actual and theoretical curve as is noted is not so much a lack of real conformity as a function of sampling error. Even large product distributions are, after all, merely samples of the total output produced over days, weeks, and months. If all this output were cast into a distribution, it

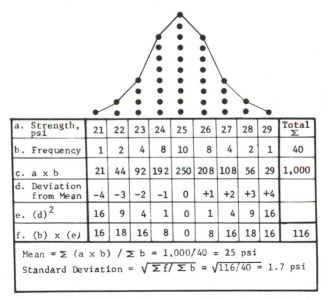

a. Strength, psi	21	22	23	24	25	26	27	28	29	Total Σ
b. Frequency	1	2	4	8	10	8	4	2	1	40
c. a x b	21	44	92	192	250	208	108	56	29	1,000
d. Deviation from Mean	-4	-3	-2	-1	0	+1	+2	+3	+4	
e. $(d)^2$	16	9	4	1	0	1	4	9	16	
f. (b) x (e)	16	18	16	8	0	8	16	18	16	116

Mean = Σ (a x b) / Σ b = 1,000/40 = 25 psi

Standard Deviation = $\sqrt{\Sigma f / \Sigma b}$ = $\sqrt{116/40}$ = 1.7 psi

Fig. 6-2. Increased variability. Compare with prior distribution pattern. The gradual spread of the distribution is reflected by larger value of standard deviation.

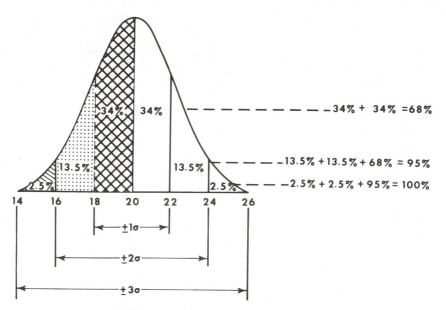

Fig. 6-3. Normal curve. Base line illustrates an application to fineness measurements, in micrograms per centimeter on a batch of man-made fibers. The average is 20 micrograms, and the percentage of product within ±1, ±2 and ±3 standard deviations around the mean is shown.

would conform much more closely to normality than even large samples.

The normal curve portrayed includes an example of fineness readings, in micrograms per centimeter, of a batch of man-made fibers. The mean is 20, and σ is 2 micrograms. Now, once the standard deviation of a normal distribution is known, the following inevitably results:

1. Approximately 68 percent of the product will fall within a range of ±1σ of the grand mean. For our example, this means that 20 ± 1σ = 20 ± (1 × 2) = 18 to 22 micrograms.
2. Approximately 95 percent of the product will fall within the mean ±2σ. Hence, for our example, a result of 20 ± (2 × 2) = 20 ± 4 = 16 to 24 is found.
3. Finally, some 99.7 percent (or practically *all*) of the product will be within a range of 3σ around the mean, which means 20 ± (3 × 2) = 20 ± 6 = 14 to 26 for our illustration.

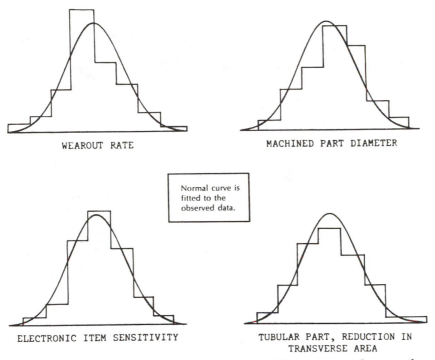

WEAROUT RATE   MACHINED PART DIAMETER

Normal curve is
fitted to the
observed data.

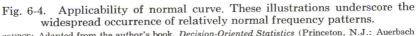

ELECTRONIC ITEM SENSITIVITY   TUBULAR PART, REDUCTION IN
                                    TRANSVERSE AREA

Fig. 6-4. Applicability of normal curve. These illustrations underscore the
widespread occurrence of relatively normal frequency patterns.

SOURCE: Adapted from the author's book, *Decision-Oriented Statistics* (Princeton, N.J.: Auerbach Publishers, 1970), with permission.

We say "approximately," because our figures are rounded. For more accurate values, and values between 1, 2, and 3 standard deviations, Table 6-1 should be consulted. Note that the first column considers the distance of σ from the mean *in only one direction* (either to the plus side or the minus side, not both). Consequently, the values 68, 95, and 99.7 percent just given should be cut in half for comparison. The remaining small discrepancies are rounding effects.

Value of Normal Curve

There are a thousand and one types of uses in which the normal curve is of considerable value in quality control and other analyses. They all spring from the fact that, given the distribution average and standard deviation, we can completely describe and evaluate it.

In practice, we usually do not know the exact values for the mean and σ, but we can estimate them with reasonable and adequate degrees of reliability. As a result, it becomes possible to estimate, from a relatively small sample, the distribution of the production lot and whether or not it conforms to specifications.

Application of Normal Curve

A cylindrical part is to be ground to a diameter of 1 ± 0.003 inch. Analysis of a number of samples gives these estimates: the process average is 1.001 and σ is 0.0025 inch. From past experience, an approximately normal distribution is assumed. Is this lot satisfactory? If not, what proportion is outside tolerance limits? Proceed as follows:

1. Determine the difference between the process average of 1.001 inches and the upper tolerance of 1.003 inches, giving 0.002 inch.
2. Form a ratio of this difference to σ. We obtain:

$$z = 0.002/0.0025 = 0.8$$

   The term z denotes the *range of standard deviation*, as used in the first column of the normal curve table.
3. From Table 6-1 for the normal curve, we find that approximately 21.2 percent of the product exceeds the upper tolerance.
4. Proceeding similarly with regard to the lower tolerance, we first find the difference $1.001 - 0.997 = 0.004$. Next, $z = 0.004 /0.0025 = 1.6$, meaning that 5.5 percent of the product will be below lower tolerance.

In practical terms, therefore, we may expect to have to regrind 20 percent of the output, and to scrap another 5 percent. The conclusion just arrived at could hardly have been anticipated or evaluated quickly without statistical methods and the normal curve.

Further Applications

What would happen now if management seeks zero scrap? For this purpose, the average setting of the grinding operation must be raised. But how much? We proceed as follows:

1. From the normal curve (Table 6-1) we note that practically all (2×49.865 percent) of the product is within $\pm 3\sigma$ of the mean.

Table 6-1.   Areas of the Normal Curve

A Range of Standard Deviation, z-value	B Percent Falling Within This Range	C Percent Falling Outside This Range
0.1	4.0	46.0
0.2	7.9	42.1
0.3	11.8	38.2
0.4	15.5	34.5
0.5	19.1	30.9
0.6	22.6	27.4
0.7	25.8	24.2
0.8	28.8	21.2
0.9	31.6	18.4
1.0	34.134	15.866
1.1	36.4	13.6
1.2	38.5	11.5
1.3	40.3	9.7
1.4	41.9	8.1
1.5	43.3	6.7
1.6	44.5	5.5
1.65	45.0	5.0
1.7	45.5	4.5
1.8	46.4	3.6
1.96	47.5	2.5
2.0	47.7	2.3
2.1	48.2	1.8
2.2	48.6	1.4
2.4	49.2	0.8
2.6	49.5	0.5
2.8	49.7	0.3
3.0	49.865	0.135
3.09	49.9	0.1

NOTES: The range in (A) is with reference to deviation to either the right side of the distribution or the left side, not with regard to both sides at one time. Critical values, most frequently applied in quality control work, are shown in full detail. In all other instances the simpler rounded terms are preferable.

Column C = Column B subtracted from 50 percent.

Example: To find the percent of pieces falling within range of 1 standard deviation on either side of the distribution average, read 34 percent (rounded) in the second column opposite 1.0 in the first column. Since under the normal curve 50 percent of the pieces lie on either side of the distribution average, 50 − 34, or 16 percent, of the pieces fall outside of a range of 1 standard deviation on either side of the distribution average, as shown in the third column.

Hence, the process average of the grinding operation must be three standard deviations above the lower tolerance of 0.997 inch.

2. Three standard deviations = 3 × 0.0025 inch = 0.0075 inch.

3. Now add 0.0075 inch to the lower tolerance of 0.997 inch. The result, 1.0045 inches, is the average setting to be aimed at.

There will now be no scrap. But how much rework will be needed? Proceed as shown below:

1. The difference between the upper tolerance of 1.003 and the process average (the new aim of production) of 1.0045 is 1.003 −1.0045 = −0.0015. Since we have crossed zero (as the minus sign of 0.0015 indicates), we know that at least 50 percent of the product needs regrinding.
2. Next, $z = -0.0015/0.0025 = -3/5$. We enter Table 6-1 without regard to the sign, except that the sign means we use column B instead of the usual column C. A value of approximately 23 percent is found.
3. Total rework expectation is now 50 + 23, or 73 percent.

We can now make cost analyses, weighing scrap against rework costs, to arrive at an optimal, lowest total cost process setting.

Third Application

Instead of adjusting the average level of the grinding operation, assume that management was able to revamp the equipment, which is now capable of performing with a standard deviation of 0.0009 inch. With the production level centered at 1.000 inch, we have, with regard to the upper tolerance:

$$z = (1.003 - 1.000)/0.0009 = 3.3$$

The normal curve table indicates that for this value of z the expectation of defectives is zero. By symmetry, the same conclusion also holds with regard to the lower tolerance. Indeed, the table stops at a z of 3.1, since for all practical purposes there is just no product beyond this distance from the mean.

Computing the Control Lines

The value of 3.1 for z might suggest that control limits should be drawn 3.1σ below the upper tolerance and 3.1σ above the lower tolerance limit. Such reasoning would be correct *if we could be sure that every sample average is the correct estimate of the process average.* But the sample average is based on inspection and testing of only a few sample pieces. Therefore, an *allowance for sampling error* must be made. Without this, we would frequently (that is, 50 percent of the time) reject perfectly good lots. This occurs merely because the sample, by chance, happens to consist of units whose average is poorer than the process average.

For specification-oriented control limits, a practical allowance permits no more than a 2- to 3-percent risk of erroneous rejection. More specifically, the risk is 2.3 percent,[1] which corresponds to the use of a z of 2.0 from the normal curve (see column C of Table 6-1). In other words, approximately 2.3 percent of the time we shall erroneously reject a good lot merely because the "luck of the draw" misrepresents actual quality. This is a theoretical risk. In practice, we should expect smaller risks, since lots are usually better than just borderline in acceptability, and sampling error decreases progressively with improved lot quality.

Next, we must consider that sample averages, being central values, exhibit less variation than the individual units of a frequency distribution. The *standard deviation of sample averages*, often known simply as the *standard error* (symbol $\sigma_{\bar{x}}$) is related to the distribution of individual units of the lot (σ). In particular,

$$\text{Standard Error} = \sigma/\sqrt{\text{Sample Size}}$$

For example, when sample size n is 4, a σ of 0.8 implies a standard error of $0.8/2 = 0.4$. We now use our previously found 3.1σ, the z-value of 2 (corresponding to a 2.3-percent risk of sampling error), and the standard error relation to write (for a lower control limit, LCL):

$$\text{LCL} = \text{Lower Tolerance} + 3.1\sigma - 2\sigma/\sqrt{n}$$

where the second term simplifies to $(3.1 - 2/\sqrt{n})\sigma$. Next, for an upper limit, we merely change one sign:

$$\text{UCL} = \text{Upper Tolerance} - (3.1 - 2/\sqrt{n})\sigma$$

Although we have a good method for calculating control limits for specification oriented control charts, the factor σ in the equation above produces undue computation labor. Fortunately, a simplifica-

---

[1]The value is justified by experience. In other words, it has been found that setting limits on the basis of $z = 2.0$ (a convenient, rounded value), yielding a risk of 2.3 percent, is a practice that has worked out satisfactorily. Those who disagree are, of course, at liberty to select a different z-value from the normal-curve table, to obtain the risk they consider acceptable.

tion can be attained by converting the formulas just given to a form in which R is used in place of σ. This will be discussed next.

Average Range in Estimating Standard Deviation

Instead of calculating σ from the squared deviations of a distribution, we may estimate it from the average range, $\overline{R}$, using the factors F_r of Table 6-2. For a sample of size 4, as an example, we obtain: Estimated $\sigma = \overline{R} \times 0.49$. Let us now return to the quantity $(3.1 - 2/\sqrt{n})\sigma$, assume that n is 4, and substitute the estimate of σ. The quantity thus becomes:

$$(3.1 - 2/\sqrt{4})0.49\overline{R} = 1.03\overline{R}$$

Had we used a nonrounded factor F_r, the answer would have been 1.02. This result checks with the factor F_s for specification-oriented control limits (in Chapter 4), for samples of size 4. All of the factors F_s were obtained by the method shown. The net result is that we can apply the readily obtainable value of $\overline{R}$ instead of having to calculate σ.

Control Limits for the Range Chart

There are two factors, D_3 and D_4, which when applied to $\overline{R}$ give control limits related to a still further type of standard deviation. This is the standard deviation of sample ranges. The resultant control limits are $\pm\ 3\sigma_{\text{range}}$ above and below the average range. While

Table 6-2.   Factors for Converting Average Range
to Estimated Standard Deviation

Sample Size	Factor* F_r	Rounded Value of F_r
2	0.8865	0.89
3	0.5907	0.59
4	0.4857	0.49
5	0.4299	0.43
6	0.3946	0.39
8	0.3512	0.35
10	0.3249	0.32
12	0.3069	0.31
15	0.2880	0.29
20	0.2677	0.27
25	0.2544	0.25

SOURCE: *Manual on Quality Control of Materials*, American Society for Testing and Materials (Philadelphia, 1951). The factors appear in Table B2, page 115 of the manual.

*These factors are more technically known as factors $(1/d_2)$. To avoid complex terminology, we substitute the term F_r.

the reasoning in obtaining range control limits is quite parallel to that for control charts for averages, the detail is too complex to be discussed here.

Operating Characteristics

Control charts, like sampling plans, have operating characteristics. For the specification-oriented charts, the curves in Fig. 6-5 are presented.

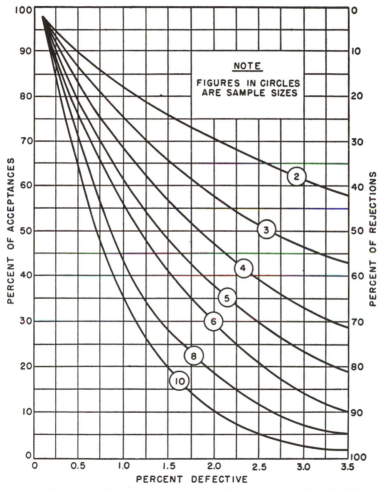

Fig. 6-5.   Operating characteristics of specification-oriented control charts.

To illustrate the use of these curves, assume that we tested six sample pieces every hour from a production process and that the output within that period contained 2.5 percent defectives. What is the probability that our control chart will call this process "out of control"? Entering the chart at the 2.5 percent defective point (base scale), we proceed vertically until the curve for sample sizes of 6 is reached. On this level, to the right, we find that this output will be rejected about 80 percent of the time. In other words, the probability of rejection is 80 percent.

Should this 80 percent be inadequate, then the sample size must be increased. As the chart shows, when n is 8, the probability of an individual sample ringing the "out-of-control" bell is close to 90 percent. For n of 10, a 2.5 percent defective run is detected about 95 percent of the time.

With small sample sizes, management may worry that the degree of control afforded may not be close enough. This reasoning, however, overlooks another important feature of all control charts: namely, that they also reveal a trend, from one plotted point to the next. Now if a series of successive sample averages show a trend toward one of the control limits, this fact is in itself an indication that something is about to go wrong with the process—and that immediate corrective action is necessary.

In practice, therefore, the really expected sampling risk should stay well below the theoretical levels of the operating characteristic curve.

Unfortunately, operating characteristic curves for range charts cannot be provided in ready-to-use form. Such curves would need to be computed individually, based on whatever process variability and specification limits are involved for a particular product. Nevertheless, although such curves are not generally referred to for range chart applications, the procedures for control of variability have been found to work well in practice.

Standard Error Effects

In the derivation of control limits for averages, use was made of the relationship that the standard error of sample averages equals the standard deviation of the original distribution, divided by the square root of the sample size. Although this effect can be proved by means of mathematical statistics only, an intuitive understanding can be gained by anyone interested. One such way, which can be readily checked by the reader, is to put the distribution of strength variations in psi (Fig. 6-2) on 40 slips of paper, each slip representing an individual test value. Record the average of successive random samples of four units, each time making sure that all slips are returned to the bowl. The resultant distribution of sample averages, superimposed on the original distribution pattern, is likely to be close to the illustration in Fig. 6-6. While sampling outcomes will vary, in *all* instances the distribution pattern of sample averages will be found to have narrowed.

In our example, the standard deviation of sample averages actually obtained is 0.87 psi. This is relatively close to the theoretical 0.85 expected by dividing the original distribution σ of 1.7 by the square root of the sample size 4. The reader may have also noted that, by keeping track of the sample ranges, he can verify the validity of the relationship $\overline{R} \times F_r$ as an estimate of σ. He need merely

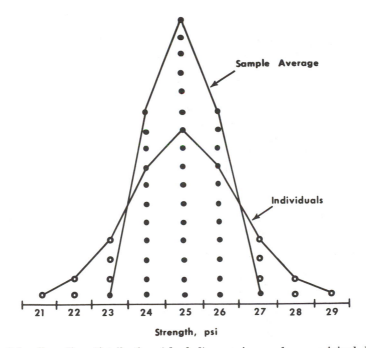

Fig. 6-6. Sampling distribution (shaded) superimposed over original distribution (of Fig. 6-2) illustrates the fact that the distribution of sample averages is narrower than that of the original, individual values.

substitute his actual $\overline{R}$ and the factor $F_r = 0.49$ for sample sizes of 4, then compare his result with the actual σ of 1.7.

Thus, while we are unable to establish mathematical proof, we can nevertheless demonstrate by means of practical sampling experiments the validity of the control chart method and the factors and relationships used.

Origin of Normal Curve

The normal curve is a function that has been derived on the basis of mathematical principles. These cannot be readily demonstrated without at least some calculus[2]. There are numerous practical situations, however, that illustrate the general way in which such a distribution comes about. As an example, take the set of possible ways in which the throw of a pair of dice may come up, in Fig. 6-7. The total of 36 combinations represents the most likely result of 36 throws, even though of course individual outcomes will vary. The distribution is beginning to shape itself along approximate normal patterns.

Imagine, next, that we were to throw three dice, then four and so on until we reach a theoretically "infinite" number of dice thrown an "infinite" number of times. At that point the distribution of possible outcomes would assume a totally smooth and perfectly normal, bell-shaped outline.

Some elementary probability calculations become apparent. For example, the odds of throwing "snake-eyes" or a two-point score are 1 out of 36 or 1/36, since only one such combination can occur out of 36 possible outcomes. Similarly, the chance of "box-cars", or two sixes, is again 1/36. Now the likelihood of throwing snake-eyes *or* box-cars is 1/36 plus 1/36, yielding 1/18. Further, the chance of a seven represents the most likely outcome, at 6 out of 36 or a probability of 1/6. Finally, two sevens in successive throws can be expected only once in 36 times, since $1/6 \times 1/6 = 1/36$.

These examples are applications of simple probability laws that serve in evaluating likely combinations and in predicting odds in games of chance, which is indeed the purpose for which the science of statistics was originated. Today we find more fruitful and constructive applications of research and science in business, industry, government, and other areas.

---

[2]See, for example, B. L. Myers and N. L. Enrick, *Statistical Functions: A Source of Practical Derivations Based on Elementary Mathematics* (Kent, Ohio: Kent State University Press, 1970).

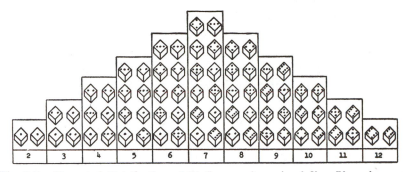

Fig. 6-7. Expected distribution of 36 throws of a pair of dice. If we increase the number of dice and the number of throws, the resultant distribution will approach more and more the bell-shaped form of the normal curve.

Origin of Standard Deviation

The nature of the standard deviation concept and the purpose for which it has been developed now become clear. We do not just calculate an index of variability, but rather a measure that has physical meaning in relation to the various areas of a normal curve. Thus, once we are dealing with a normal distribution, calculation of the standard deviation gives us full information regarding the over-all range of the distribution and the proportion of items in the distribution that will fall within a given range of standard deviations.

An interesting further use of the standard deviation can be made when we express it as a percentage of the distribution average, thereby obtaining a value known as *coefficient of variation, v*. For example, we previously (in Fig. 6-2) showed an approximately normal distribution with a mean of 25 psi and a standard deviation of 1.70 psi. We now have:

$$\text{Variation Coefficient, } v = 100 \times 1.70/25$$
$$= 6.8 \text{ percent}$$

A conversion from absolute values (pounds per square inch in this example) to relative and hence widely comparable terms of percent has been accomplished. It is for this reason that occasionally the variation coefficient is referred to as the "relative standard deviation." But the value of v goes farther, particularly when comparisons among different variations are to be made.

Standard Deviation and Variation Coefficient

In the following we shall show how the variation coefficient may at times yield more comparable and thereby more meaningful results than the standard deviation. A manufacturer of insulating sheets was able to maintain the following characteristics of two of his products:

	Style A, Heavy	Style B, Light
Average weight, ounces per yard	15	10
Standard deviation, ounces per yard	0.3	0.3

From an examination of standard deviations it would appear that the plant has been equally successful in maintaining weight control, which is an important factor from both a cost and quality viewpoint. Yet, when coefficients of variation are computed, a different picture is revealed:

	Style A, Heavy	Style B, Light
Variation Coefficient, percent	2	3

It is thus shown that the *relative variability* of Style A was less than Style B. The above is a simplified version of an actual situation in which not just two but a dozen major styles were compared. Several inconsistencies in relative variability were uncovered, which in turn led to the identification and correction of the causes in processing (setting of machinery; blending consistency of materials; control of feeds, speeds, and tensions) and to a better product at lower materials costs.

The principle brought out is that in all instances where product variations may be expected to increase proportionately with increases in the average of the distribution, the variation coefficient may be a superior, comparative measure. In other situations, however, the variation coefficient may be misleading, particularly in many types of dimensional control problems. For example, the problem of machining a certain part to a basic dimension of, say, 15 inches is usually no different from machining another part to a size of 10 inches. Therefore, the standard deviation alone should be used to express variability. Sometimes, faulty usage may pervade an entire industry. In the construction industry, for example, the author found a predominant practice of expressing strength variability of certain types of products in terms of variation coefficient. Yet, when a careful analysis was made, it was discovered that product variability did not tend to increase with the average level of strength. In the absence of such a proportionate, consistent relationship, the use of standard deviation seems preferable.

Summary

We have covered a range of concepts, and it may be well to take stock. First of all, the standard deviation was developed as a measure of variability. Its relation to another, less precise but nevertheless practically useful measure, the range, was shown. Next, the use of standard deviation values in developing underlying distributions, by means of the normal curve, was brought out. Sample averages vary less than individual observations, but the standard deviation of sample averages (also known as the standard error) was shown to be related to the standard deviation of individuals. Finally, the distribution of production lots often tends to follow at least an approximately normal pattern. Gathering up the relationships brought out, we can now combine the effects of product tolerances, production lot distributions, measures of variability in individuals and averages, and the characteristics of the normal curve to develop control limits for sample averages. The limits provide a set of criteria for the purpose of indicating instances of out-of-control or off-standard output. We have also given the common-sense explanation of several of the concepts used (in lieu of mathematical derivations), and we have given attention to the relation of the standard deviation and the variation coefficient.

Some readers may find it necessary to go over this chapter more than once to comprehend its full meaning. It should be emphasized, however, that for practical applications a broad level of understanding is all that will be needed. A return to detailed reading may thus be deferred to the reader's convenience without hindering him in the use of subsequent materials in this book.

REVIEW QUESTIONS

1. How are standard deviation and variance related?
2. Assume that you know the mean and standard deviation of a produc-

tion process. Then, what percent of the product will fall within a range of the mean and (A) + 1 standard deviation, (B) + 2 standard deviations, (C) + 3 standard deviations, (D) ± 2 standard deviations? Assume a normal distribution.

3. What is the value of tables of areas under the normal curve?
4. How are average range and standard deviation related?
5. What is the standard error?
6. How are standard deviation and variation coefficient related?
7. In order to assure minimum tensile strength of 15 kilograms for a plied filament yarn, with a process standard deviation of 0.5 kilogram, what is the lowest level of the process average that can be tolerated?

In Appendix 3, Cases 2, 3, and 8 may be further analyzed by preparing frequency distribution plots as a useful supplement to control chart data.

CHAPTER 7

Further Control Charts

In this chapter, a number of additional, highly useful control charts will be described. We shall examine in turn: control charts computed from center lines, charts for percent defective, the chart for number of defects, and the defects-per-unit chart.

CHARTS FROM CENTER LINES

In industry many situations arise for which it is important to have control charts, even though no tolerance limits have been specified. In those instances, control limits can be calculated with reference to the so-called "center line." The latter is the desired or specified average level of the production process.

Calculation

We set control limits by use of the A_2 factors provided in Table 7-1. The formulas are:

$$\text{UCL} = \text{Process Average} + (\overline{R} \times A_2) \text{ and}$$
$$\text{LCL} = \text{Process Average} - (\overline{R} \times A_2)$$

The limits occur three standard errors above and below the center line. We refer to them, most commonly, as "3-sigma limits." They involve a negligible chance (0.1 to 0.2 percent, as reference to normal curve values shows) of erroneous rejection of good production. But they are so wide as to introduce a certain lack of sensitivity in early detection of unsatisfactory processes. Accordingly, some people prefer 2-sigma limits. For this purpose, one need merely multiply A_2 by the ratio 2/3.

The term A_2 is composed of the elements $3 \times F_r / \sqrt{n}$. The factor 3 was just explained. F_r converts $\overline{R}$ to σ and the subsequent division by the square root of the sample size n changes the standard deviation of individual units to the standard error, reflecting variations in sample averages.

87

Table 7-1. Control Limit Factors for Charts for Averages
Based on Center Lines

(a)	(b)	(c)	(d)
Sample Size	Factor A_2	Rounded Value A_2	$(2/3) \times A_2$
2	1.880	1.88	1.253
3	1.023	1.02	0.682
4	0.729	0.73	0.486
5	0.577	0.58	0.385
6	0.483	0.48	0.322
8	0.373	0.37	0.249
10	0.308	0.31	0.205
12	0.266	0.27	0.177
15	0.223	0.22	0.149
20	0.180	0.18	0.120
25	0.153	0.15	0.102

SOURCE: American Society for Testing and Materials, *Manual of Quality Control of Materials* (Philadelphia, Pa., 1951), Table B2, p. 115.

RESULT: Factors A_2 yield control limits that are three standard errors above and below the center line (actual or desired process average).

NOTE: For control limits two standard errors above and below center line, multiply A_2 by the ratio 2/3. While ordinary A_2 factors involve a risk of error of 0.3 percent of erroneous rejection of good lots, $(2/3) \times A_2$ gives risk of error of 2.5 percent for *either* an upper limit *or* a lower limit (thus, a joint error of 2.5 + 2.5 = 5 percent).

Control of Color and Flavor

Colorimeter tests on a cola drink reflect the consistency of both color and composition of flavoring ingredients. From successive samples of size 3, a process average of 225.3 and an average range of 5.5 were obtained (see Table 7-2). The factor A_2 for sample size 3 is 1.02. Hence, 3-sigma limits are:

Control Limits $225.3 \pm (5.5 \times 1.02) = 219.7$ to 230.9

Management desired more sensitive 2-sigma limits, however. We therefore multiply 1.02 by ⅔ and substitute the resultant 0.68, giving:

Control Limits $= 225.3 \pm 5.5 \times 0.68 = 221.6$ to 229.0

The actual use of these limits is now illustrated in Fig. 7-1. One out-of-control point, ascribable to excessive attenuation of the flavoring compound, known as "dope," was noted and corrected.

Control of Processing Time

Our next illustration, in Fig. 7-2, is concerned with the control of

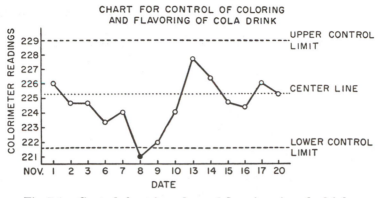

Fig. 7-1.   Control chart for color and flavoring of a cola drink.

fulling time, which is a finishing operation in the preparation of dyed woolen cloth. It should take no more than 75 to 80 minutes to "full" a fabric and give it the desired finished characteristics. Sometimes, however, more time is needed to accomplish requisite quality because of errors in prior processing stages or because settings, admixtures, and finishing solutions in the fulling operation are faulty.

Control data are obtained by means of randomly spaced visits to the fulling machines. While it is relatively simple to check the amount of fulling time required to achieve satisfactory quality (with the aid of an elapsed time recorder on each machine), it is more difficult to establish instances where operator carelessness results in inadequate fulling time. Other tests, however, on finished fabric do give indications of the occurrence of insufficient fulling.

The charting is done for both averages and ranges. One instance of out-of-control variability was ascribable to an inexperienced operator, who did too much fulling on some units and not enough on others. With additional help and instructions, this source of quality variation was corrected. The charts are of particular interest, since they reflect a frequent occurrence in production: quality and efficiency of output are often correlated. When one is controlled, the other tends to follow in a similar direction.

Multiple Control Charts

It will be noted that in many types of processing a variety of quality characteristics all have equal importance. Consequently,

Table 7-2.   Control Data for Cola Drink Daily Colorimeter Readings on Bottling Line No. 1

Day	Bottle No. 1	2	3	Sample Average (Rounded Out)	Sample Range
1st	222	228	229	226	7
2nd	223	225	231	226	8
3rd	225	221	223	223	4
4th	222	224	226	224	4
5th	228	226	230	228	4
6th	222	226	224	224	4
7th	228	226	224	226	4
8th	220	218	228	222	10
9th	230	224	225	226	6
10th	224	220	228	224	8
11th	230	224	227	227	6
12th	224	222	226	224	4
13th	227	228	226	227	2
14th	224	222	225	224	3
15th	225	231	227	228	6
16th	226	220	228	225	8
17th	223	227	229	226	6
18th	228	226	224	226	4
19th	226	224	228	226	4
20th	221	229	221	224	8

Total   4506   110

Average 225.3   5.5

control charts reflecting several test series may have to be maintained.

For example, the producer of bottled drinks is concerned with more than just colorimeter values. Specific gravity and degree of carbonation are of equal importance. Moreover, he cannot be sure of flavoring quality until a panel of tasters has checked small amounts of "blind" samples. If the tasters, from these samples, can identify a current batch as different from a reference standard, then an out-of-control condition exists. By the term "blind" we refer to the fact that the taster is not permitted to know the identity of samples

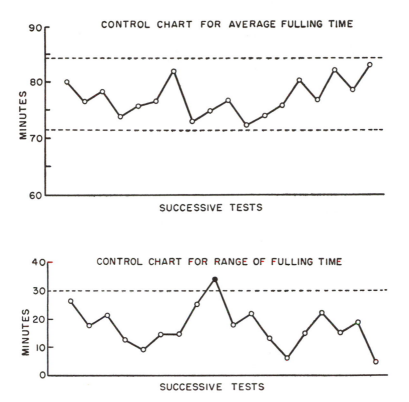

Fig. 7-2. Control chart for processing conditions. Illustration refers to processing time in a finishing operation, involving the fulling of a dyed fabric.

tasted until after he has made his evaluation. Control limits are needed to segregate chance from real variations in tasters' findings.

Multiple control charts may be kept separately, or else they may be placed in comparative form, as illustrated in Fig. 7-3. This chart is concerned with the chemical analysis of steel used in tire molds. Because the molds are cast in plain surfaces and the treads are machined afterward, free-machining sulphur steel is used. Charts are maintained on the following components:

1. *Carbon.* Must be sufficient to prevent tearing of the metal in front of the tool bit, but should not be too high in order to avoid excessive hardness and costly machining.
2. *Sulphur.* Must be sufficient to promote easy machining, yet as low as possible to save processing costs.

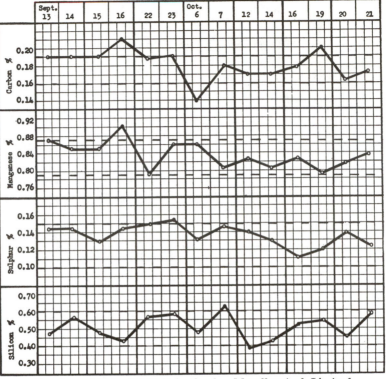

Courtesy of Robert J. Feltrin, Quebec Metallurgical, Limited.

Fig. 7-3. Multiple control chart. Illustration of sulphur steel chemical analysis.

3. *Manganese.* Requires close control for reasons similar to carbon and sulphur.
4. *Silicon.* Must be adequate to avoid gassy castings, but low enough to prevent undue deoxidation. The latter would increase hardness and machining costs.

We note that relatively narrow control limits were accomplished, as a result of low process variability. While some out-of-control conditions did occur, they were generally corrected quickly.

Group Control Chart

When several machines, which are practically alike, process the same material and output, a single group control chart may replace a whole set of individual graphs. The following is involved:

1. The chart contains an upper and lower control limit.
2. Each hour, day, or other time period, only two values are plotted. These are (A) the highest and (B) the lowest average for the group.
3. The machine number is recorded in connection with each plotted point.
4. When a machine number keeps on appearing at either the high or low level of the chart, it serves as an indication that off-standard output is impending for that machine.
5. When an out-of-control point appears, that machine and, where indicated, all other machines are checked.

An illustrative example for a four-spindle winding machine is given in Fig. 7-4. The out-of-control point for Lot No. 105 might have been prevented, had management paid closer attention to the repeated high (though within control) average for spindle No. 2.

Many types of manufacturing can be noted where the number of machines, number of output positions (such as spindles), or other aspects of operation involve large groups on essentially the same operations. A consideration of group control charts in such situations then becomes imperative.

CHARTS FOR PERCENT DEFECTIVE

All of the charts described thus far have involved test data recorded in terms of scalar values, such as pounds, inches, degrees,

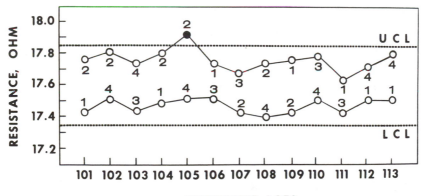

Fig. 7-4. Group control chart. Example of four-spindle winder. Spindle numbers are indicated. Resistance of wound spools is the quality characteristic of interest.

and the like. Often, however, it is either impossible, impractical, or inconvenient to record more than an "attribute" of an inspected workpiece or other item. The attribute may be "good" or "defective," "conforming or nonconforming," "on-standard" or "off-standard," "go" or "not-go," and so forth. In evaluating the quality of a lot under these conditions, all we can do is to note the percentage of product that has a certain attribute. For example, if 5 percent of a sample is "defective," then 100 — 5 percent or 95 percent is "good," excepting that the term "effective" is often used in place of the less specific "good."

Records of Percent Defective

In order to function properly, a hydraulic component must meet exacting requirements for quality of finish, as checked by comparison with a reference standard. Samples of 200 from successive lots yielded the number of defectives and the percent of defectives shown in Table 7-3. A "defective" is, of course, a unit that does not exhibit requisite surface finish. The total of the percentages is 28. With ten lots recorded, this results in an average *percent defective*, $\overline{p}$ percent of 2.8. We could also say, as some do, that the (decimal) *fraction defective* averages 0.028.

The percent defective, when subtracted from 100 percent, yields *percent effective*, which is 97.2 in our example, denoted by the symbol $\overline{q}$ percent. The value of $\overline{p}$ also represents our process average.

Standard Deviation of a Percentage

We are now ready to calculate the standard deviation of a percentage, σ_p. The formula is different from that for sample averages based on scalar values or "variables." In particular,

$$\sigma_p = \sqrt{(\overline{p}\% \times \overline{q}\%) \div n}$$

where n is the familiar term for sample size. Now substitute $\overline{p} = 2.8$, $\overline{q} = 97.2$ percent and $n = 200$, except that for simplicity of calculations we shall convert the percentages to decimals:

$$\sigma_p = \sqrt{(0.028 \times .972)/200}$$
$$= 0.0117 \text{ or } 1.17 \text{ percent.}$$

The determination of the standard deviation is thus readily accomplished. Since it is based on samples, it represents an estimate of the true but unknown standard deviation of the process.

Table 7-3.   Record of Percent Defective
(Inspection Results of Quality of Finish of a Hydraulic Component)

Lot No.	Defectives in Sample of 200	Percent Defective
1	4	2
2	8	4
3	2	1
4	6	3
5	4	2
6	4	2
7	10	5
8	6	3
9	4	2
10	8	4
—	—	—
Total	56	28
Average	5.6	2.8

Control Limits for Percentages

It is customary to establish control limits at levels of three standard deviations above and below the process average, in percentages. Hence:

$$\text{Control Limits} = \text{Process Average} \pm 3\sigma_p$$

For a standard deviation of 1.17, therefore, $3\sigma_p$ equals 3×1.17 or 3.5 rounded. Added to the process average of 2.8 percent, this yields an upper control limit of 6.3 percent. Subtracting 3.5 from the process average gives a negative result. The lower control limit is therefore taken as zero.

Using Percentage Charts

From the data gathered above, the control chart with the appropriate upper limit is drawn, as shown in Fig. 7-5. By plotting inspection results for each lot, one notes that all of the output falls inside the limit. The process is thus under good control.

Where the chart shows a number of out-of-control samples, an investigation of processing factors is called for. The first aim of such a check would be to assure that inspection and testing are done

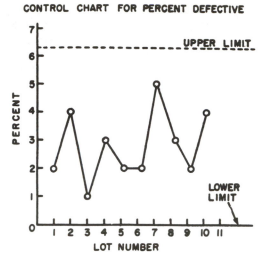

Fig. 7-5. Control chart for the quality of finish of a hydraulic component.
Process is in control.

properly, consistently, and uniformly. Once this is established, the
method of sampling should be reviewed to eliminate the possibility
of nonrandom or otherwise nonrepresentative procedures. As a
third and demanding step, the processing must be investigated,
both as regards general setups and operating procedures and the
particular occasions when off-standard results are noted on the control chart.

Purpose of Lower Limits

A lower control limit may be well above zero. Now, since zero
defects would be the ideal attainable, why should one worry if a
sample falls below the lower control limit? The reasons for this will
become apparent when we consider why and how such an occurrence may develop:

1. The luck of the draw in sampling may produce an erroneous
   value below the lower limit. Such an event should happen
   rarely. Indeed, with 3-sigma control limits, it should happen
   no more than once or twice in a thousand samplings.
2. It may represent careless or superficial inspection or failure of
   the inspector to record the actual defectives in his sample
   properly.

3. A true lowering of defective output may have somehow been accomplished.

The last item bears further discussion. Sometimes a temporary or inadvertent change in process settings may result in unexpected quality improvements. We want to recognize this condition as soon as it happens, so that we can consider the advisability of making the "inadvertent" occurrence a standard practice. At other times, a low percentage of defective product may reflect unduly low production speeds or raw materials inputs that are too costly. Insidious changes may occur. For example, in one manufacturing organization there were gradual changes in raw stock specifications over a period of years, resulting in improved processing conditions but also raising costs to a level where the competitive position of the firm and the marketability of its products suffered.

A familiar story begins to repeat itself: he who controls quality also helps control costs. These are not just theoretical pronouncements, but are findings that will be confirmed in every organization where an effective control program is in operation.

Control of Processing Conditions

The percent defective chart may also be applied to control processing conditions. A typical example is taken from the spinning of a man-made fibers yarn. The spinning room contains twelve machines (called "frames" in the industry), each equipped with 100 spindles. At any one time, one or two machines may be down for maintenance, changeover, or other causes. On each of the running machines, a number of spindles may be unproductive or "idle" for one of several reasons: mechanical defects, runout of supply that had not been replaced as yet, a yarn break that had not been pieced-up by the operator, and the like.

It is important to control the level of idle spindles so that corrective measures can be taken when the percentage exceeds an allowable limit. A common practice is to make daily checks, by walking through the room and recording observations, such as shown in Table 7-4. From experience, a value of 1 percent idle is considered standard. We shall use this value as the center line (in lieu of the process average, in percent). Using the average number of spindles of 1,000 on running frames as the sample size, and $100 - 1 = 99$

Table 7-4.   Control of Machine Efficiency

Date	No. of Machines Running	No. of Spdls. on Running Machines	No. of Spindles Idle	Percent Idle
2/1	9	900	12	1.3
2	10	1,000	12	1.2
4	8	800	6	0.8
5	11	1,100	11	1.0
6	10	1,000	10	1.0
7	10	1,000	15	1.5
8	12	1,200	24	2.0
9	10	1,000	10	1.0
11	10	1,000	10	1.0
12	10	1,000	8	0.8
Average	10	1,000		

percent as the "percent effective" or percent productive spindles, we find:

$$\sigma_p = \sqrt{1\% \times 99\% \div 1,000}$$
$$= \sqrt{0.01 \times 0.99 \div 1,000}$$
$$= 0.003 \text{ or } 0.3\%$$

Next, the control limits become:

$$\text{Control Limits} = 1\% \pm 3 \times 0.3\% = 0.1 \text{ to } 1.9\%$$

The lower limit can be ignored, since it is so small. The upper limit is drawn in on the control chart as shown in Fig. 7-6.

A Modification of the Percentage Chart

Instead of recording percent defective, it is sometimes preferred to utilize a chart showing "number of defects in the sample." Such a chart is often more readily understood by production people. In our example of hydraulic components, for example, we could have said "defectives found in each sample of 200" in place of percent defective. The upper control limit, instead of being 6.3 percent, then becomes 6.3 percent × 200 or 12.6 defectives. In actual practice this means that if up to 12 defectives are found in a sample of 200, the production process is still considered in control. A number of 13 or greater then signals an out-of-control status.

This type of chart is based on the assumption that the sample size

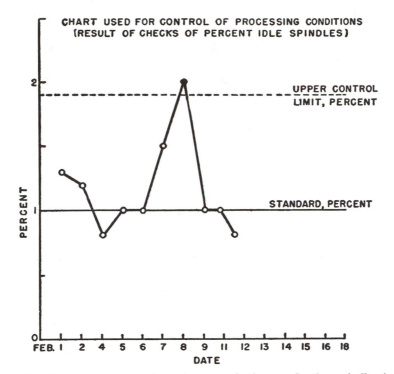

CHART USED FOR CONTROL OF PROCESSING CONDITIONS
(RESULT OF CHECKS OF PERCENT IDLE SPINDLES)

Fig. 7-6. Percent defective chart for control of unproductive spindles in a yarn spinning department. Note out-of-control point on February 8.

will always be constant. In the mechanical hydraulic parts illustration, this requirement is fulfilled. However, in the case of idle spindles control, the sample size varies somewhat. On one day, 800 spindles were checked, while on others 1,200 were examined, depending on the number of machines in operation. Now let us assume that we found 10 idle spindles in one day. This represents 1.25 percent of 800 but only 0.83 percent of 1,200 spindles. Accordingly, a chart for "number of idle spindles" would have been misleading to some extent.

It may be further pointed out that, even for a percent defective chart, excessive fluctuations in sample size cannot be ignored. After all, n does enter into the formula for the standard deviation. Since it appears under a square root, however, its effect is relatively moderate. So long as sample sizes vary no more than ± 20 percent, no problem should be encountered. When there are greater fluctuations, control limits must be constantly recomputed, based on the effect of n on the standard deviation. The procedure becomes quite cumbersome, and there are fortunately only rare instances in which it must be used.

DEFECTS-PER-UNIT CHARTS

The defects-per-unit chart applies primarily in connection with visual examination of large, bulky, or continuous products, where quality is measured in "defects per unit." One unit may thus have several defects. For example, a refrigerator may have a faulty handle and a warped panel, thus two defects. A yard of fabric may contain a rust spot and a slub, thus two defects. In each instance, we have two defects per unit, even though at one time the unit is the refrigerator and at the other time it is the yard of fabric.

Now, a yard may be broken down into smaller subdivisions, such as a tenth of a yard. Such small lengths are unlikely to ever contain more than one defect, and we can then talk about the percentage of 1/10th yard lengths (out of the total of tenth yards in the fabric) that contain defects. We have thus converted the data to percent defective. But this is not always a feasible procedure. For example, we cannot break a refrigerator down into equal small segments, such as we did with the fabric. The product itself is not of a nature to lend itself to meaningful, equal-sized segmentation.

The defects-per-unit chart thus has its place in the control of product quality.

Standard Deviation of Defects-per-Unit

The formula for the standard deviation of defects-per-unit, σ_c, is one of engaging simplicity:

$$\sigma_c = \sqrt{\text{Average Number of Defects per Sample}}$$

The average is akin to a center-line value. It may represent the average of a series of sample values, or it may represent a desirable and feasible standard. In most practical production situations, it may be wise to begin with an actual process average, based on experience, and to shift to a standard only when assurance has been received (from accumulated data) that a center-line value or other standard is routinely attainable.

The short-cut symbol for average number of defects per sample is a $\bar{c}$ (c-bar), and hence the chart is often known as a *c-chart*.

Defects-per-Unit Control Limit

By custom, 3-sigma limits are used. Thus we have:

$$\text{Control Limit} = \bar{c} \pm 3 \sqrt{\bar{c}}$$

For an illustration, we shall consider a plastic coated fabric which is 100 percent inspected after finishing. Nevertheless, to assure the proficiency of this screening, a recheck sampling procedure is in use. From each lot of 200 rolls, 10 are selected randomly, and 50 yards per roll are checked. The sample is thus 500 yards.

From past experience, the number of defects missed by screening has been found to average 4 for a good, conscientious screening inspector. Letting $\bar{c}$ be 4, then, we have:

$$\text{Control Limit} = 4 \pm 3 \sqrt{4} = 0 \text{ to } 10$$

In other words, if recheck inspection uncovers more than 10 defects per 500 yards, missed by the original inspector, this fact indicates lack of proficiency or proper care in that operation.

Defects-per-Unit Chart in Production

For an example of the defects-per-unit chart in the continuous control of production, we shall use the same bottling plant for which we previously set up colorimeter control. This time it is concerned with 100-percent inspection of the finished, bottled product. (See Fig. 7-7.)

Each bottle passes behind an illuminated magnifying glass. Any foreign matter—such as cigarette ends, tin foil, and other items, thoughtlessly inserted into empties by the public—must be noted. Although there is machine scouring of each bottle, some items nevertheless do not wash out. The examiner's job is to find and remove these and otherwise faulty bottles.

At the end of each day, the quality control department receives a small truck, containing several cases of rejected bottles, each marked as to the production line responsible. From this, it is a simple matter to compute the number of rejected bottles or "rejects" per shift for each line. Over a period of time, 16 rejects per shift are the average number normally attainable with the plant's

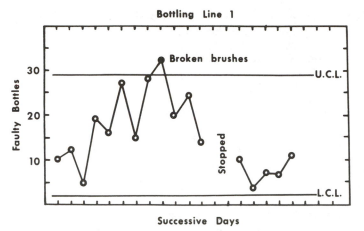

Fig. 7-7.   Control chart for bottling line inspection.

scouring equipment. Control chart limits are then readily computed. In particular:

$$\text{Control Limits} = 16 \pm 3 \sqrt{16} = 4 \text{ to } 28$$

In practical use, an excess of actual rejects over control limits furnishes a good reason to investigate the scouring operation for wear of brushes, improper settings, and other possible causes. If, on the other hand, the rejects fall below lower control values, we may be interested in interchanging inspectors on different lines. If the "high quality" and low rejects follow the inspector, we would have good reason to question the proficiency of inspection.

GENERAL NOTES

This concludes our discussion of further control charts. The reader may wonder why we are not presenting the operating characteristics of these charts. Like the range-chart, all of the charts presented here do not lend themselves to the preparation of ready-made curves. Common industrial practice is to use these charts without calculating individual operating characteristic curves.

It should be noted that all test data in scalar form can be converted to percentages. For example, for our Rockwell Hardness of shovel blades application (Chapter 4), the tolerances were 42 and 48. Hence, instead of recording actual hardness numbers, we could have simply said that any value within 42 to 48 is "good" and any value outside is "bad." Translation to "effective" and "defective" then permits a "percent defective" control chart. Such, indeed, is possible. But in doing so, we have thrown away valuable information about the degree of goodness or badness of each unit. This loss of detail means that, in general, a percent defective control chart requires from 10 to 15 times the sample size of a variables chart (one that maintains the scalar measurements) in order to be of equal efficiency (have approximately the same steepness of the operating characteristic curve). One should thus go slowly in substituting a percent-defective chart.

A word about the distribution of sampling results. For the percent-defective chart, the calculation of the standard deviation is based on the assumption of a so-called *binomial distribution*. This use is *always* justified, no matter what the distribution of the lot may be, *so long as sampling is at random*. In technical language, we say that the percent-defective control chart method is "nonparametric," that is, it does not depend on the parameter representing the lot distribution. In a similar, practically parallel manner, the standard deviation of the defects-per-unit data is based on a so-called *Poisson* distribution, which is again always justified under random sampling.

REVIEW QUESTIONS

1. What are the characteristics of control charts computed from center lines? How do they differ from charts computed from tolerance limits?
2. What kind of chart would you recommend to control the amount of maintenance and repair time consumed in each of five processing departments? (Give an example of how this chart would look.)
3. The head of a shipping department is concerned about the number of errors made by clerks in filling orders (shortages, overages, wrong items, faulty addressing, incorrect postage). What chart should be maintained to control, on a weekly basis, the percentage of orders handled incorrectly?
4. For the shipping department just discussed, what kind of chart should be maintained to control the number of customer complaints per 1,000 orders filled?
5. A manufacturer must maintain close control over the number of defectives in a metal blanking operation. He has the choice of using lot-by-lot sampling inspection or a control chart for percent defective. Both methods are valid. Which one would you prefer, and why?
6. Dust count in a white room, where sensitive equipment is assembled, must not exceed 10 particles per cubic foot per hour. What kind of control chart would you apply?

7. Refer to the data of question 8, Chapter 4, this time assuming that no tolerance has been established yet. Nevertheless, we wish to assure that processing stays as close to the optimal dimension of 60 millimeters as is practically possible. A control chart based on center lines is required. What are the upper and lower limits of such a chart, based on (A) rounded factor A_2 and (B) factor $2A_2/3$?

8. For the drop hammer data in Chapter 2, what is (A) the process average, (B) the upper control limit, and (C) the lower control limit?

9. Inspection and testing of successive lots of 25 printed circuit boards yielded these results:

Lot No.	101	102	103	104	105	106	107	108
No. of defective boards per lot	2	0	1	0	1	1	0	3
No. of minor defects per lot	4	2	0	8	0	1	2	3

Using control chart criteria (control limits for percent defective boards and for number of defects per lot), are any lots out of control; and if so, which?

10. Tensile tests of a nylon filament produced the values below in kilograms per skein:

Lot No.	101	102	103	104	105	106	107	108
Test No. 1	21	22	20	18	21	18	24	20
Test No. 2	19	24	21	16	22	18	20	22

Based on control charts for sample averages and ranges, are there any out-of-control lots? Which, if any?

Cases 5 and 6 in Appendix 3 provide practice in setting control limits for percent defective and for defects per unit. For Cases 2, 3, and 8, if it is assumed that no tolerance limits have been established, control charts based on center lines can be computed.

CHAPTER 8

Process Capability and Product Tolerances

Quality cannot be inspected into a product. It must be created as part of product design and built into the components during production. It follows that control charts cannot be used with success unless the production process is capable of fulfilling the specifications and tolerances envisioned by the engineering design. Only when process capability and product tolerances are compatible with each other and with the functional performance requirements of the finished product, can inspection and quality control perform a useful role.

Neither are tolerances a purely technological consideration. Every specification and each specification limit will carry deep implications regarding materials costs, manufacturing speeds, processing expenditures, and final price. These factors, in turn, must be evaluated in relation to the market for a product.

In a well-coordinated quality control program, the inspection and testing data obtained from surveillance of production are fed back to the design group, so that specifications and tolerances can be considered and the design and redesign of products made consistent with process capabilities, performance requirements, and marketing considerations.

In the following, we shall examine important aspects of product specifications and tolerances, methods of evaluating process capability in relation to these factors, and ways and means of successful and economic reconciliation of product tolerances with product specifications.

SPECIFICATIONS AND TOLERANCES

From the design of a product to the ultimate consumer stretches a chain of requirements for usefulness, economy, and quality—forming the basis of what are known technically as *specifications*.

The user of a razor blade, for example, asks merely that it give him good shaves at low cost. To meet his needs, the designer must establish specifications for dimensions, thickness, centrality of perforations, temper, sharpness, evenness, and finish. These in turn set the pattern for the manufacturing operations of blanking, perforating, heat-treating, grinding, polishing, and finishing. There must also be specifications for the quality of the steel, such as microstructure and composition, which constitute a set of standards for the steel mill. Further, for certain processing operations, such as plating, a set of detailed instructions with regard to methods and procedures will be necessary.

Specifications are thus the specific instructions established to maintain quality standards. Since there is variability in materials, processes, methods, and products, each specification must provide (at least implicitly) a tolerance, delineating a tidemark between allowable and unacceptable amounts of variation around the specification average.

Effect of Production Variables on Tolerances

Realistic tolerances which are neither too close nor too wide are often difficult to set. The engineer who designs a new product and establishes specifications and tolerances for it often has no way of predicting with accuracy the effect of the many variables of production on the uniformity of the article. Even if a particular new item is produced on known equipment, there may be differences in the material, the tools and dies, the shape and dimensions, or the requirements for closeness of fits and clearances. As a result, it often becomes necessary to change tolerances, and sometimes even the entire design, after an item has been in production for some time and experience with it has been gathered.

Tolerances may be either too small or too large. The costliness of such excesses in either direction is obvious. If tolerances are set too large, the product will be unsatisfactory. If tolerances are too small, the plant will incur the high cost of maintaining precision greater than really needed to make a good article. Machines and equipment may have to be stopped too frequently for realigning and readjustment, and product is rejected needlessly. The information obtained by systematic inspection can often be of invaluable aid in attaining proper tolerances. How to achieve this objective will now be discussed.

When Tolerances are Too Large

The fact that tolerances are set too large may be established by inspection or possibly by the functioning of the completed product. Let us assume that component parts have been carefully checked after each machining operation and that practically no defective product has been passed. Yet when the parts are to be assembled, some of them do not fit together properly. Matching, reworking, and discarding of parts become necessary. If the pieces are found to measure within the tolerances, then the obvious conclusion is that the tolerances are too large.

Tolerances for Nonprecision Products

The importance of tolerances is often not realized where a nonprecision product is manufactured. For example, in an air-vise assembly of a spring steel blank with a plastic handle the only specifications maintained were those for the dimensions of the pre-drilled hole and the small metal tang that was to be forced into it. Despite careful setting of the air-vise and selection of only skilled operators, a high proportion of defective assemblies—loose, cracked, and bent handles—occurred. Tolerances were reduced, and still no improvement resulted. Carelessness, improper electronic preheating of the handles, and excessive polymerization of the plastic were blamed, until finally the chief source of defective work was discovered—varying length of the handle. Little attention had been paid to the length of the handle, since whether a tool had a handle of 4 inches or $4\frac{1}{8}$ inches seemed unimportant in practical use. But it did make a difference in the way the handle fitted into the vise. If the fixture was adjusted for 4 inches, but the handle happened to measure $4\frac{1}{8}$ inches in length, the metal blank would be pushed just $\frac{1}{8}$ inch deeper into the plastic than permissible and thus crack the handle. Conversely, a shorter handle would give a loose assembly. The problem was solved by establishing specifications of $4'' \pm \frac{1}{32}$ for the handle length and performing careful process inspection on the handle-shaping operation. Any future defective assemblies could be definitely traced to improper setting of the fixture, insufficient or excessive preheating of the handle prior to assembly, or carelessness.

When Tolerances are Too Small

Tolerances which are smaller than necessary may be as costly as those which are too large. They occur frequently because of the design engineer's understandable fear of "sticking his neck out." If he sets too large a tolerance, it will ultimately show up in parts which do not assemble readily with proper fits and clearances, electrical characteristics which do not fulfill requirements, or in some other way. On the other hand, small tolerances may present practical production problems and greatly increase the cost of manufacture, but they will not show up as glaring mistakes in the form of defective product.

It is not surprising, therefore, that tolerances on blueprints tend to be very conservative, whereas the quantity-minded production people tend toward more liberal tolerances, and the quality-minded inspector is caught in-between these opposing views. The confusion resulting from unrealistic tolerances need not be elaborated.

Inspection data can be used to advantage in a shop seriously concerned with the setting and maintaining of realistic specifications and tolerances which can be met under mass manufacturing conditions and which will give a satisfactory end product.

PROCESS CAPABILITY

In order to evaluate effectively the capability of a production process, it is first desirable that the process be operating in a so-called "state of statistical control." This means that control charts for averages and ranges have been maintained on the production process for some time and that only an occasional out-of-control occurrence has been observed. The chart for averages is preferably a chart involving 3-sigma limits computed from the center line. If and when these conditions are fulfilled, then the American Society for Quality Control defines process capability as *the least variability of quality that a process is capable of maintaining.*[1]

Evaluating Process Capability

When a control chart for averages and ranges has been placed on a process and due attention has been paid to the control limit and any trends of plotted test results, a fair to good degree of control should soon be attainable. The average range can then be used in finding the standard deviation, and from this the capability of the process can be evaluated.

In the filling of 10-cc. vials of an injectable pharmaceutical, for

---

[1]Standard A3-1694, *Glossary of Terms Used in Quality Control,* Milwaukee, Wisconsin: American Society for Quality Control (1964).

example, the following data served in the calculation of process capability: (1) The process was known, from past experience, to have an approximately normal frequency pattern; (2) the average range $\overline{R}$ based on the last 20 controlled-process-state samples, was 0.4 cc. Sample size was 25. Then:

$$\text{Standard Deviation of Process} = \overline{R} \times F_r$$
$$= 0.4 \times 0.25 = 0.1 \text{ cc.}$$

Here F_r is the tabulated value, for samples of size 25, for converting $\overline{R}$ to σ. Now, since we know from the normal curve that practically all of the product of a distribution falls within a range of $\pm 3\sigma$ of the distribution center, we have:

$$\text{Process Capability} = \pm 3\sigma = \pm(3 \times 0.1) = \pm 0.3 \text{ cc.}$$

Some people will say that the *natural tolerance* of the process is ± 0.3 cc. or simply 0.6 cc. The term "natural tolerance" carries the further implication that the process is capable of maintaining a given range, such as 0.6 cc. for some time during ordinary operation, without the action of any person in making process adjustments. But these are fine and often difficult-to-interpret distinctions. For example, how long is "some time," and what is the difference between an operator's giving "normal attention" or "making an adjustment"?

Note that this approach of finding process capability, or, as some say, the *limits of process capability*, is valid only for controlled processes from relatively normal distributions.

Capability Without Control Charts

Situations may arise where we must deal with processes on which control charts have not yet been established. Or else, such charts were developed but somehow we have been unable, thus far, to attain a state of statistical control. Such processes are not really "uncontrolled," but they do lack the degree of control that would be needed if the average range $\overline{R}$ is to be used in evaluating capability.

Fortunately, there are other tools available to us. One of these is the preparation of a frequency distribution pattern. For this purpose, it is desirable to utilize from 100 to 200 individual test results or more, spread over a sufficient period of time to reflect the true range of variation of the process. Next, proceed as follows:

1. Observe the maximum width of the distribution. For example,

the refractive index of 200 glass beads, used in industrial processing, may yield a highest value of 2.5 and a lowest value of 1.7. The maximum width is therefore 0.8. We may also refer to this as the *maximum range*.

2. Calculate the standard deviation of the distribution. For the 200 beads, assume a standard deviation of 0.2.

3. Find the *expected maximum range*. From the normal distribution, applied to the glass beads, this expected maximum range is $\pm 3\sigma$ os simply 6σ, hence $6 \times 0.2 = 0.12$.

4. Compare the observed maximum width with the expected maximum range. The larger value is the process capability. In our example, comparing the maximum width of 0.8 with the expected maximum range of 0.12, we find that the process capability is 0.12 or ± 0.6.

The reasoning behind these procedures should be obvious. Statistical expectation, based on the assumption of a normal curve, predicts that practically all of the product will fall within a range of 6σ or $\pm 3\sigma$ around the center of the distribution. But our distribution may not be normal to either a slight or marked degree. Conservative procedure, then, is to take the larger of the two values of maximum observed width versus expected maximum range.

Practical Adjustments

In many manufacturing organizations it is recognized that process capability is a type of minimal variation that may not be realized at all times in production operations. Such manufacturers will then make some gross adjustment in the value of the capability, which is then considered a safety factor.

For the vial-filling operation, for example, the limits of process capability were found to be ± 0.3 cc. If management now makes an allowance of, say, 33.3 percent for "safety factor," the adjusted limits become ± 0.4 cc. No objection is raised to such practice, provided it is based on past experience supported by ample objective data and analyses. The tragedy, in many situations, is that little rational basis exists for the safety factor used, and as a result the limits used in practice are too wide. This nonstatistical approach, it will be demonstrated, can be quite costly not only in terms of manufacturing expenditures but also in terms of bidding on contracts or entering markets where tight specifications must be (and could be) met.

TOLERANCE-CAPABILITY INTERACTIONS

Although we have previously considered the ways in which the interaction between product tolerances and process capability affects quality and cost of output, we shall now give this topic further attention. Our analysis will be enriched by the fact that process capability is now established on the basis of statistical studies.

Capability and Single Tolerance Limit

To examine the interaction of process capability with a single tolerance limit, we shall recall the illustration of a vial-filling operation. We are required to provide at least 10 cc. in each vial. The limits of process capability were found to be ±0.3 or simply 0.6 cc.

If management is to satisfy the lower limit of the tolerance, the process average must be far enough from this point to assure that practically no product is below 10 cc. From the nature of the normal curve, this means that we must center the filling operation 3 standard deviations above 10 cc. Since 3σ corresponds to 0.3 cc (one half of the capability limit, as stated above), we have:

$$\text{Required Process Average} = \text{At Least: 10 cc.} + 0.3 \text{ cc.}$$
$$= \text{At Least 10.3 cc.}$$

Depending on the amount of safety margin that is desired, we can go to higher averages. But if we set the safety margin too wide, we shall be needlessly supplying vial contents above the amount needed, and for which we are not paid.

We may try to reduce process variability, so that the new process capability is reduced to, say, ±0.22 cc. The process average can now be set closer to the lower tolerance, at a level of 10.22 cc., as against the prior 10.3. Figure 8-1 portrays these relationships. The saving of 0.08 cc. is only 0.08 cc. per vial, or 0.8 percent of total cost of vial content. Over the long run, however, this will become a sizable gain.

In almost all instances where process capability and tolerances have not been evaluated, either process quality improvements or production and materials cost savings are likely to be potential values. We have given an illustration of a filling operation, of which there are numerous applications throughout industry. Bottling of beverages, stuffing of candy, filling of cosmetics jars, canning of food, and baling of fabric are some examples. Overfilling occurs whenever the product contains larger amounts or proportions of an

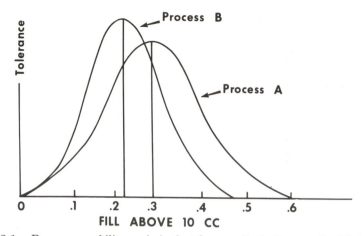

Fig. 8-1. Process capability and single tolerance limit. Process B with lower variability has an average closer to the lower limit of fill, thus saving costly materials. If the capability of process A can be similarly improved, it, too, can be moved closer to the lower limit of fill.

ingredient than really required. Often, moreover, the excess ends up as waste. In the manufacture of pile rugs, for example, yarn loops are produced which are then trimmed at the top. Undue variability or excessively high loops all mean the same: that part or portion of the yarn that extends beyond the intended depth of pile is trimmed off as waste.

Knowledge of process variability, process capability, and tolerance requirements is needed in order to (1) press for realistically attainable process variability in production, (2) set averages in relation to tolerances and process capability, (3) attain quality at a relatively minimum expenditure of time and cost.

Capability and Two-sided Limits

When an upper *or* a lower specification limit only is involved, an appropriate shift in the average level of the process will always accomplish this goal. Not so for two-sided tolerances, where a shift of the average in either direction may merely switch off-standard output from one side of the specification to the other.

As an illustration of two-sided limits, let us reconsider our earlier hardness testing application on shovel blades. The tolerance was set at ±3 around a required average of 45 hardness numbers. Since a normal distribution spread is ±3σ, it follows that a distribution with σ = 1.0 for hardness values will just meet tolerances. Actually, of course, we prefer a somewhat smaller standard deviation than 1.0, so as to leave a safety factor. An excess of σ beyond 1.0 leads to off-standard product, as pattern B of Fig. 8-2 illustrates.

Assume now that the actual process standard deviation is indeed less than the maximum of 1.0 allowable, at 0.95 in Rockwell numbers. Then 45 ± 3 × 0.95 gives limits of 45 ± 2.85, or from 42.15 to 47.85, which is slightly below the tolerance. This situation is reflected by pattern A of Fig. 8-2. The safety factor is relatively small. A more conservative approach would be to seek a safety factor of 20 percent, if this is economically attainable, in which case processing should be held to a standard deviation of 0.8.

However, what would be the desirable action if process standard deviation is quite readily maintained at, say, 0.5? The difference between the tolerable maximum of 1.0 and 0.5 yields a 50-percent safety factor.

Economic considerations will now come into the foreground. For example, if materials costs or processing time can be saved by moving the process average closer to the lower tolerance, this can

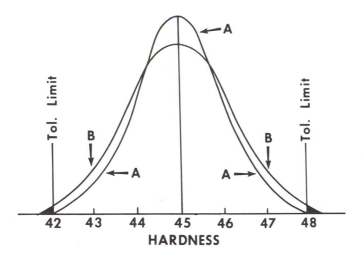

Fig. 8-2. Two-sided tolerance and process capability. When variability is excessive, process B, off-standard product occurs. Process A barely satisfies the tolerances, but leaves little margin for error.

now be accomplished. Assuming that a 20-percent safety factor is still desired, we would proceed in this way:

1. Multiply the standard deviation by 1.2. Thus, $0.5 \times 1.2 = 0.6$. The standard deviation has now been enlarged 20 percent to take account of the desired safety margin.
2. Since one-half of a normal distribution is encompassed by 3σ, multiply 0.6 by 3, yielding 1.8.
3. Add 1.8 to the lower tolerance limit of 42. The resultant 43.8 is the new, desired process average.

The analysis of one-sided and two-sided limits in relation to process capability serves to reemphasize the desirability of maintaining process variability at economically narrow levels.

Tolerances and Control Limits

If tolerances are not compatible with process capability, then control chart applications are likely to be inadequate and to yield unsatisfactory results. This may be recognized from a review of the factors involved in setting control limits.

The specification-oriented control chart is obtained by moving from the tolerance limits toward the center line. Now, if tolerances are too narrow, the control limits will likewise be too close. Management will find an excessive number of sample averages that fall above and below the control lines. The cause responsible may be (1) process variability higher than the process is capable of, or (2) tolerances that have been set unrealistically close, when the capability of the process is considered. Sometimes management may find itself in a bind. The inability of the process to satisfy close tolerances has been established, but the functional and contractual requirements of the product or assembly do not permit a widening of the tolerances. In such instances, resort to 100-percent screening must be had, to sort out the nonconforming portion of each day's output.

When control charts are based on center lines, no automatic tie-in with tolerances is provided. The calculations are independent. As a result, it is not uncommon to observe instances in industry where a manufacturer believes his process to be operating well, while actually it is merely staying within control limits that are too wide from a viewpoint of tolerance requirements. It is for these reasons that this author does not recommend center-based control lines in instances where tolerance limits on the individual unit of product must be met.

How is it possible that some production processes continue to turn out nonconforming product without a "day of retribution"? Several factors account for this. Often it turns out that the unit tolerance has been set too tightly. The designer was overly conservative in his demands, and no assembly failures may occur, or else the functional and assembly failures that do happen are not frequent enough to cause concern. In other instances, where components are shipped to others (customers or affiliated plants), the receiving inspection is so inadequate as not to detect off-standard conditions. In many other cases, bad product merely slips through, giving rise to customer dissatisfaction and field complaints. The false sense of security, created by the erroneous use of control charts based on center lines, can become a particularly insidious factor in delaying corrective action.

Systems Aspects of Tolerances

From a review of the statistical, technical, and managerial aspects of tolerances; process capability; process variability; and control limits; one should be impressed by the manner in which interlocking factors must be carefully weighed in arriving at proper values and decisions. Together, the factors noted form a closely woven system. We must analyze the relationships in order to understand what is happening. Next, upon a consideration of quality, performance, and economic values involved, proper decisions regarding processing, tolerances, and control methods must be developed. Although we have demonstrated the relationships involved and given tools for analysis and control, not all of the sophisticated aspects have as yet been covered. The further considerations involved will be discussed in Chapters 14 and 15.

TOLERANCE SYSTEMS IN INTERCHANGEABLE MANUFACTURE

A large proportion of statistical quality control applications occur in the machine-building and metalworking industries. Now, while the principles of statistical methods are equally valid for all industries, there are certain aspects of "interchangeable manufacture" that occur in the machine-building and metalworking industries which should be understood before one begins quality control installations.

The principles to be discussed in the following have more general implications. For example, the assembly of electronic equipment also requires compatibility and interchangeability of components. Related concepts apply to the blending, mixing, attenuation, and other processing of chemical constituents. As far afield as the textile industry, we can find need for interchangeable compatibility. For example, the weaving of a fabric is nothing but the assembly of yarn, following a prearranged design, in which the various yarns must conform to standards and tolerances for diameter and density.

Advantages of Interchangeability

By "interchangeable manufacture" we refer to products that are formed of parts made to the degree of accuracy required for the proper functioning of any part in any one of a number of duplicate machines or devices, even when the parts are selected at random and regardless of the lot to which they belong. Strict interchangeability, however, may not always be feasible. Under practical working conditions, therefore, there are modifications of this system or different degrees of interchangeability.

The interchangeable manufacturing system serves to produce duplicate mating parts sufficiently accurate or uniform in size to insure assembling (at random) and proper functioning without fitting individual parts together. In some cases each part may be made to fit any mating part in a limited series, but this is not true interchangeability. In other cases, parts are assembled without fitting but not until they have been selected so that each mating pair has the desired fit. Thus, companion or mating parts which happen to be made to the extreme limits of the allowable size are not supposed to interchange, in many cases, if assembled at random; consequently, *"selective assembly"* is necessary. To illustrate, a shaft of maximum diameter may not assemble in a hole of minimum diameter, and such a combination, or one close to it, is likely to occur whenever parts are assembled at random. However, by selective assembly suitable fits may be obtained. This method is a modification of interchangeable manufacture. Most mechanical devices that are supposed to be interchangeable have some parts which are strictly interchangeable and others which must be selected to obtain proper fittings.

The chief advantage of interchangeable manufacture is in the increase in manufacturing efficiency and resulting reduction in manufacturing costs. When duplicate parts of a mechanical device are sufficiently accurate to permit interchangeability, the ease with which broken parts may be replaced is another advantage, but is decidedly secondary when compared with the reduction in costs.

Determining Degrees of Accuracy
Required for Different Machine Parts

In this mechanical age, machines and other mechanical appliances are used for so many different purposes that great accuracy is required in producing some of them and comparatively little accuracy in producing others. Then, too, different classes or qualities of fits are needed even between parts of the same device to insure proper functioning. For instance, some assembled mating parts may require very tight fits, others, fits which are not tight but are close enough to eliminate play or clearance, and still others, fits with a certain amount of clearance. The fundamental reason for such variations relates to the general purpose of the machine or other device, its quality, and the function of each part or group of parts entering into its construction. To illustrate the matter of function, a shaft in one case must revolve freely in a bearing but possibly with no appreciable amount of play. In another case, play is not objectionable. Then there are very tight fits, as illustrated by a plug or pin which is somewhat larger than the hole to receive it, so that sometimes tons of pressure are required to obtain a rigid and fixed assembly of the parts. Numerous other variations might be cited, all of which are related in the last analysis to the function of the part or parts. This means that the fits between assembled parts must be based either upon data established for similar applications or upon experience and judgment until more definite information is available. The question of fits leads to a very important point which will now be considered.

General Method of Specifying Fits
of the Required Quality

When a given class or quality of fit is known to be satisfactory, the next step is to provide means of obtaining it in a systematic way in the normal course of manufacturing practice. To take a concrete illustration, suppose that a certain fit is required between a cylindrical pin and a hole. It is evident that obtaining the desired fit necessitates some method of controlling the accuracy of both pin and hole. This leads us to the general and very important subject of *tolerances*.

First, we are confronted by the fact that absolute accuracy is impossible. Second, the closest possible approach to it would be very costly, and third, in machine-building practice varying degrees of inaccuracy may be allowed or tolerated without impairing the function or practical usefulness of the machine or of the individual parts which form it. For example, suppose that experience with a given machine indicates that a certain shaft having a nominal diameter of 1 inch will function properly with a diameter variation anywhere, say, between 1 inch and 0.995 inch. Permitting this allowable error, or tolerance, of 0.005 inch instead of aiming for perfection may mean the difference between economical manufacturing practice and an utterly impractical condition. Therefore, in economical manufacturing the following steps are essential:

1. Certain classes of fits between mating parts of a mechanical device are established.
2. The kind or quality of fit is based upon proper functioning of the mating parts and of the entire mechanism. The durability of the device, as well as its initial performance, usually would be an important factor in determining what constitutes proper functioning.
3. The relation between a given quality of fit and proper functioning may have been established by repeated performances of the same or similar apparatus. If such tested-fit data are lacking, then experience and judgment must be relied upon because the tolerances and resulting allowances which determine fits cannot be established by employing fixed rules or formulas. In case the machine is an entirely new type, designers must rely as far as possible upon their experiences with other jobs that are closest in their approach to the one at hand.
4. The different parts of a machine, especially if quite a number are required in its construction, have different classes of fits and degrees of accuracy because of the various functions.
5. In machine-building practice, the range of allowable fits, and also of allowable errors in manufacturing, is quite large because mechanical devices in general vary from those which must be very precise to those which may be crude and inaccurate without impairing their usefulness.

6. Fits even for work intended to be of the same class or quality must vary to some extent because variations are inherent in all manufacturing processes; but by controlling them so as to avoid errors too large to permit proper functioning of the part or parts, great manufacturing economy is effected.

7. The quality of a fit and the permissible variation depend upon fixing whatever maximum and minimum limiting dimensions will give the desired quality, but since the dimensions of acceptable parts vary and may lie anywhere between the extreme limits allowed, the fits vary accordingly as, for example, when the fits are between pins and holes or other internal or external parts.

8. After the tolerance has been established for a given part, thus fixing certain minimum and maximum limits to the overall dimensions, means should be provided in the regular course of manufacture to maintain these limiting sizes. To this end, control charts, with limits based upon both the tolerances for a given part and the variability of the production process, are finding increasing application because of their efficiency and economy.

How Variability of Product Affects Fits Between Mating Parts

Since small dimensional variations within the allowable limits are to be expected, it is evident that there will be corresponding variations in the tightness or looseness of fits between mating parts such as the fit of a shaft in the hub of a gear or pulley. To illustrate, if a shaft of the maximum allowable diameter should happen to be assembled in a hole of the minimum allowable diameter, the tightest fit would be obtained. If these conditions were reversed and the minimum shaft were assembled in a maximum hole, then the loosest fit would be obtained. Suppose as an example of fit variation, we take the case of a cylindrical part having a maximum diameter of 2 inches and a minimum diameter of 1.9995 inches. Assume that this part is to fit into a hole having a minimum diameter of 2 inches and a maximum diameter of 2.0008 inches. In this case, the maximum shaft and minimum hole are the same size, but if a minimum shaft happens to be assembled in a maximum hole there will be a clearance of 0.0013 inch. One might wonder if such minimum or maximum fit values would always be acceptable. The answer is *no* in many cases where unusual precision or uniformity of the fittings is essential. This point will be considered later.

Basic Dimensions

A basic dimension or size may be defined as a theoretical size to which a tolerance is applied in order to obtain a degree of accuracy which meets practical requirements or is within *allowable* minimum and maximum limits. Now when maximum and minimum limiting dimensions have thus been established for some part, the basic size may either be the minimum or the maximum and it may also, in some cases, be between (usually midway) the maximum and minimum limiting dimensions. If the basic size is the minimum, then the tolerance is plus relative to it; if the basic size is maximum, the tolerance is minus relative to it; and if the basic size lies between the maximum and minimum dimensions, the tolerance is divided— equally as a rule—so that it is both plus and minus relative to basic.

If the basic diameter of some cylindrical part is, say, 2 inches and the tolerance is $+0.0000$ and -0.0005, then the allowable range of sizes is from the maximum diameter of 2 inches down to the minimum diameter of $2 - 0.0005 = 1.9995$ inches. If the hole for this cylindrical part has a basic size of 2 inches and the tolerance is $+0.0008$ and -0.0000, the allowable range of hole sizes is from 2 up to 2.0008 inches. As a general rule, the minimum diameter of a hole is its basic size. In the case of a mating shaft, the maximum diameter usually is the basic size.

Before dealing further with the relation between basic dimensions and tolerances, it might be well to mention that the American National Standard for Preferred Limits and Fits for Cylindrical Parts includes a *preferred series* of basic sizes or diameters (Table 8-1). The object of a preferred series of basic sizes is to reduce the number of diameters commonly used in a given size range.

Relation Between Tolerances and Basic Dimensions

If all of the tolerance is in either a plus or a minus direction relative to a basic dimension, the tolerance is known as "uniteral" because it is in one direction. For example, if a shaft diameter is 2 inches $+ 0.0000 - 0.0005$, the tolerance of $- 0.0005$ is minus and unilateral. If the tolerance is divided relative to a basic dimension, it is known as "bilateral." For example, if the center-to-center distance between two holes is given as $6 + 0.003 - 0.003$ or 6 ± 0.003 inches, this indicates that the tolerance may be either plus or minus 0.003 inch, and it is known as "bilateral."

Unilateral tolerances are recommended usually for mating sur-

Table 8-1.   American National Standard Preferred Basic Sizes

Decimal			Fractional					
0.010	2.00	8.50	1/64	0.015625	2 1/4	2.2500	9 1/2	9.5000
0.012	2.20	9.00	1/32	0.03125	2 1/2	2.5000	10	10.0000
0.016	2.40	9.50	1/16	0.0625	2 3/4	2.7500	10 1/2	10.5000
0.020	2.60	10.00	3/32	0.09375	3	3.0000	11	11.0000
0.025	2.80	10.50	1/8	0.1250	3 1/4	3.2500	11 1/2	11.5000
0.032	3.00	11.00	5/32	0.15625	3 1/2	3.5000	12	12.0000
0.040	3.20	11.50	3/16	0.1875	3 3/4	3.7500	12 1/2	12.5000
0.05	3.40	12.00	1/4	0.2500	4	4.0000	13	13.0000
0.06	3.60	12.50	5/16	0.3125	4 1/4	4.2500	13 1/2	13.5000
0.08	3.80	13.00	3/8	0.3750	4 1/2	4.5000	14	14.0000
0.10	4.00	13.50	7/16	0.4375	4 3/4	4.7500	14 1/2	14.5000
0.12	4.20	14.00	1/2	0.5000	5	5.0000	15	15.0000
0.16	4.40	14.50	9/16	0.5625	5 1/4	5.2500	15 1/2	15.5000
0.20	4.60	15.00	5/8	0.6250	5 1/2	5.5000	16	16.0000
0.24	4.80	15.50	11/16	0.6875	5 3/4	5.7500	16 1/2	16.5000
0.30	5.00	16.00	3/4	0.7500	6	6.0000	17	17.0000
0.40	5.20	16.50	7/8	0.8750	6 1/2	6.5000	17 1/2	17.5000
0.50	5.40	17.00	1	1.0000	7	7.0000	18	18.0000
0.60	5.60	17.50	1 1/4	1.2500	7 1/2	7.5000	18 1/2	18.5000
0.80	5.80	18.00	1 1/2	1.5000	8	8.0000	19	19.0000
1.00	6.00	18.50	1 3/4	1.7500	8 1/2	8.5000	19 1/2	19.5000
1.20	6.50	19.00	2	2.0000	9	9.0000	20	20.0000
1.40	7.00	19.50	.....	....	...	......	....	....
1.60	7.50	20.00						
1.80	8.00	.....		All dimensions are given in inches.				

faces such as fits between internal and external cylindrical or other parts. However, when the parts are tapering in form, the tolerance may be unilateral in some cases and bilateral in others. If the variation in the position of, say, a taper plug in a hole, is less likely to cause trouble in one direction than in the other, then the tolerance should be unilateral and either plus or minus. If a minus tolerance will tend to locate the plug away from the preferred direction, obviously the tolerance should be plus. If a variation in either direction is equally objectionable, then the tolerance should be divided and bilateral. The center-to-center distances between holes usually are bilateral because ordinarily a variation in either direction is equally objectionable. The center-to-center distance between the bearings for gears, however, represents one exception. In this case, the tolerance should be unilateral and plus because a slight increase in the center-to-center distance is not likely to cause trouble; but a decrease in center distance might result in excessive pressure between the teeth and cause unsatisfactory operation.

Relation Between Tolerance and Allowance

Tolerance or allowable error should not be confused with *allowance*, which indicates the difference between the dimensions of mating parts. An allowance is intended to provide whatever fit or degree of tightness or looseness may be required between the assembled parts. If the internal member, such as a cylindrical pin or shaft, is somewhat smaller than the external member, such as the hub of a lever, pulley or gear, then the allowance represents *clearance*. On the contrary, if the internal member is larger than the external part, the allowance represents *interference of metal*; consequently, the parts must be assembled either (1) forcibly, as in a hydraulic press; (2) by heating the outer member and thus expanding it enough to permit assembly; or (3) by cooling the inner member by means of dry ice or liquid nitrogen, thus contracting it enough to permit assembly.

The actual allowance is affected by the actual tolerances and resulting sizes of both hole and shaft. To illustrate, assume that a pin has a maximum size of 4 inches and a tolerance of +0.0000 —0.0006. If we assume that the hole to receive this pin has a minimum size of 4 inches and a tolerance of +0.0010 — 0.0000, the minimum allowance would be zero and the maximum allowance 0.0016 inch. These variations in allowances and resulting fits, which are due to the fact that the actual tolerances on mating parts may vary from zero up to the maximum, occur in connection with both internal and external members. Consequently, allowances between assembled parts may vary considerably, and the extremes occur when a maximum pin and minimum hole or minimum pin and maximum hole are assembled together.

Now, if the need for tolerances could be entirely eliminated and if every part could be made to an exact dimension, it is evident that allowances for fits could be absolutely uniform. However, tolerances are essential; hence, where variations in fit allowances must be exceptionally small, there are two possible methods of procedure. One method is to reduce the tolerances, but if the reductions are excessive it may not be possible to utilize standard manufacturing equipment, or, in fact, any kind that would not result in prohibitive manufacturing cost. A second method of securing tolerance uniformity and often the only economical one, is to produce the parts efficiently—without attempting to secure such exceptional accuracy that special methods are required, and then to select or sort the parts so that extreme size and allowance variations are avoided. This method of "selective assembly," as it is called, is to insure assembling those parts which, as a result of selection, are known to have the required allowance. This selection or sorting of groups according to size is sometimes done automatically by utilizing automatic gages or gaging machines which are now available in various forms. Automatic gaging may be employed either to detect parts which are not within the maximum and minimum limits, or the apparatus may be designed for the more complete operation of sorting parts into a number of groups according to their sizes, as in selective assembly. The latter method is commonly employed in size grouping such parts as antifriction bearing balls or rollers.

Standardization of Tolerances and Allowances

The great importance of tolerances and allowances in the manufacture of mechanical devices has resulted in the adoption of standards for the general guidance of those who specify these dimensional values (these standards will be found in engineering handbooks). Previously, there was reference to American National Standard Preferred Basic Sizes. Since preferred tolerances and allowances are a logical complement to preferred sizes, a series of tolerances and allowances has been included in the American National Standard. This series is listed in Table 8-2. The total range, as the table shows, is from 0.0001 to 0.2500 inch. The purpose of this series is to reduce, as far as possible, the total number of different tolerances needed to meet practical requirements. In this Standard series there are only fifty-one tolerances between 0.0001 and 0.2500 inch. When the engineer or designer specifies tolerances taken from whatever part of this Standard series meets his requirements, he avoids using a lot of tolerance values which are in between and which therefore differ from the Standard series but have no practical advantage over this series.

Table 8-2.   American National Standard Preferred
Series of Tolerances and Allowances*

0.1	1	10	100	0.3	3	30	...
...	1.2	12	125	...	3.5	35	...
0.15	1.4	14	...	0.4	4	40	...
...	1.6	16	160	...	4.5	45	...
...	1.8	18	...	0.5	5	50	...
0.2	2	20	200	0.6	6	60	...
...	2.2	22	...	0.7	7	70	...
0.25	2.5	25	250	0.8	8	80	...
...	2.8	28	...	0.9	9	...	...

*Data in Thousandths of an Inch

The American National Standard for Preferred Limits and Fits for Cylindrical Parts provides 10 grades of fits, as shown in Table 8-3, so arranged that for any one grade similar production difficulties are encountered in maintaining tolerances for a given range of sizes. Appropriate tolerances for holes and shafts can be readily selected. It should be noted that the tolerances for various classes of fits were designed on a unilateral hole basis.

The tolerance grades, in turn, have a definite relation to the types of machining operations that will under ordinary conditions produce work of acceptable quality, as shown in Fig. 8-3.

Classes of Fits for Cylindrical Parts

The American National Standard for preferred limits and fits for cylindrical parts provides a number of classes of fits ranging from running and sliding to force and shrink fits. These fits, designated as RC, LC, LT, LN, or FN, provide standard limits for hole and shaft for a series of size ranges extending, in some cases, from 0.04 to 200 inches. The limits are selected so that the fit obtained by mating parts in any one class will produce approximately similar performance throughout the size ranges.

The following classes of running and sliding fits are provided: Class RC 1 close sliding fits are intended for accurate location of parts which must assemble without perceptible play; Class RC 2 sliding fits intended for accurate location but with greater maximum clearance than Class RC 1; Class RC 3 precision running fits

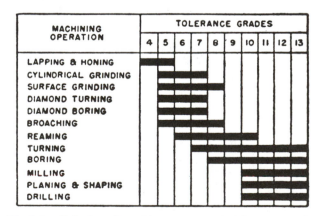

Fig. 8-3.   Relation of machine processes to tolerance grades.

Table 8-3. American National Standard for Preferred
Limits and Fits for Cylindrical Parts

Nominal Size, Inches		Grade									
Over To	4	5	6	7	8	9	10	11	12	13
	Tolerances in thousandths of an inch*										
0- 0.12	0.12	0.15	0.25	0.4	0.6	1.0	1.6	2.5	4	6
0.12- 0.24	0.15	0.20	0.3	0.5	0.7	1.2	1.8	3.0	5	7
0.24- 0.40	0.15	0.25	0.4	0.6	0.9	1.4	2.2	3.5	6	9
0.40- 0.71	0.2	0.3	0.4	0.7	1.0	1.6	2.8	4.0	7	10
0.71- 1.19	0.25	0.4	0.5	0.8	1.2	2.0	3.5	5.0	8	12
1.19- 1.97	0.3	0.4	0.6	1.0	1.6	2.5	4.0	6	10	16
1.97- 3.15	0.3	0.5	0.7	1.2	1.8	3.0	4.5	7	12	18
3.15- 4.73	0.4	0.6	0.9	1.4	2.2	3.5	5	9	14	22
4.73- 7.09	0.5	0.7	1.0	1.6	2.5	4.0	6	10	16	25
7.09- 9.85	0.6	0.8	1.2	1.8	2.8	4.5	7	12	18	28
9.85- 12.41	0.6	0.9	1.2	2.0	3.0	5.0	8	12	20	30
12.41- 15.75	0.7	1.0	1.4	2.2	3.5	6	9	14	22	35
15.75- 19.69	0.8	1.0	1.6	2.5	4	6	10	16	25	40
19.69- 30.09	0.9	1.2	2.0	3	5	8	12	20	30	50
30.09- 41.49	1.0	1.6	2.5	4	6	10	16	25	40	60
41.49- 56.19	1.2	2.0	3	5	8	12	20	30	50	80
56.19- 76.39	1.6	2.5	4	6	10	16	25	40	60	100
76.39-100.9	2.0	3	5	8	12	20	30	50	80	125
100.9 -131.9	2.5	4	6	10	16	25	40	60	100	160
131.9 -171.9	3	5	8	12	20	30	50	80	125	200
171.9 -200	4	6	10	16	25	40	60	100	160	250

* All tolerances above heavy line are in accordance with American-British-Canadian (ABC) agreements.

are about the closest fits which can be expected to run freely (these are intended for precision work at slow speeds and light journal pressures); Class RC 4 close running fits are intended chiefly for running fits on accurate machinery with moderate surface speeds and journal pressures, where accurate location and minimum play are desired; Classes RC 5 and RC 6 medium running fits are for higher running speeds, or heavy journal pressures, or both; Class RC 7 free running fits are intended for use where accuracy is not essential, or where large temperature variations are likely to be encountered; and Classes RC 8 and RC 9 loose running fits are intended for use where material such as cold-rolled shafting and tubing made to commercial tolerances are involved.

Locational fits include: Class LC locational clearance fits intended for parts which are normally stationary but which can be freely assembled or disassembled; Class LT locational transition fits which are a compromise between clearance and interference fits, for application where accuracy of location is important, but either a small amount of clearance or interference is permissible; and Class LN locational interference fits which are used where accuracy of location is of prime importance and where parts require rigidity and alignment with no special need for bore pressure.

Force fits include: Class FN 1 light drive fits which require light assembly pressures and produce more or less permanent assemblies; Class FN 2 medium drive fits which are suitable for ordinary steel parts or for shrink fits on light sections. Class FN 3 heavy drive fits which are suitable for heavier steel parts or for shrink fits in medium sections; and Classes FN 4 and FN 5 force fits which are suitable for parts which can be highly stressed or for shrink fits where the heavy pressing forces required are impractical.

Unified and American National Standard for Screw Threads

This important standard provides several classes of external and internal screw threads which may be combined in various ways to obtain a number of different fits. Thus, there are Classes 1A, 2A, and 3A external threads and Classes 1B, 2B, and 3B internal threads, Classes 1A and 1B have the widest tolerances and Classes 3A and 3B the narrowest. In addition, the older Classes 2 and 3 for both external and internal threads have been retained from the former American Standard. Although most frequently a Class 1A external thread will be used with a Class 1B internal thread, a Class 2A external thread with a Class 2B internal thread, and so on; actually any class of external thread—1A, 2A, 3A, 2, or 3—may be combined with any class of internal thread—1B, 2B, 3B, 2, or 3—to obtain the kind of fit desired.

Since interchangeable manufacture is applied extensively in mass production, its economic importance is apparent.

CAPABILITY AND METROLOGY

The accuracy and precision going into the quality measuring process affect the determination of product capability, the setting of tolerances, and the establishment of control limits. Unfortunately, the subject of metrology—the science of measures and weights—is so formidable and complex that a proper treatment would call for a large volume. In the following we shall confine ourselves to certain essential aspects of the topic, which are basic prerequisites for anyone applying quality control methods. Primarily, therefore, the discussion will be concerned with principles, and the vehicle of illustration will be the dimensional aspects of metrology.

Dimensional Metrology

Control of dimensions of product, such as width, length, and diameter, requires the use of many types of measuring equipment. The simplest instrument for this purpose is the steel rule, which permits fairly reliable determinations to within one-hundredth of an inch. This accuracy is usually not sufficient, and in order to attain greater precision, calipers are generally used. A common caliper, the micrometer, permits direct readings up to one-thousandth of an inch, with readings to one-ten-thousandths when equipped with a vernier scale. For still greater precision, dial gages, precision gage blocks, electronic gages, compressed air, and optical devices may be used.

When large quantities of products are to be inspected, it is of advantage to use gages that have been built or adjusted to the limits required. This approach permits quick checks of product for conformance to the necessary dimensions. Considerable ingenuity has been developed by engineers in designing gages for checking various classes of dimensional errors.

Limit Gages

Various types of gages must be employed to check the linear dimensions, angular characteristics, amount of taper, roundness, concentricity, parallelism, and contour—as, for example, the involute form of a gear tooth. Thus, interchangeable manufacturing depends not only upon tolerances, but also upon precise and efficient measuring and gaging instruments for ascertaining that product conforms to the tolerances that have been established.

It is not within the scope of this treatise to describe and illustrate the important types of gages and gaging machines, or to present various automated and semiautomated instruments for measuring and gaging with the aid of compressed air, electronic sensors, or optical equipment. While these devices bring a high degree of reliability, precision, and resolution to the measuring process, they do not alter the basic principles of gaging. It will thus be adequate for our purpose to refer to a few limit gages of basic design, partly to illustrate the relation between limits and primarily to show the general principles of limit gaging.

The limit gage is a very common type and, as the name implies, it shows whether the size of a part is within the allowable maximum and minimum dimensions; but the fixed limit gages do not show the actual size. One form of limit gage which is commonly used for external measurements is shown at *A*, Fig. 8-4. This gage and the other examples of gage design here illustrated conform to the American Gage Design Standard. The gage illustrated at *A* has two sets or pairs of measuring pins. If the work is within the allowable limits, it will pass between the first pair but not between the second pair. This explains why a gage of this general type is often referred to as "go" and "not-go" or as "go" and "no-go." Model *C* is shown at the right in Fig. 8-4. It has two flanged gaging pins or "buttons" and a single block anvil opposite.

At *A* in Fig. 8-5 is shown a common type of limit gage for holes. The "go" end at the left is longer than the "not-go" end at the right so that the user can distinguish readily between the two. The gage illustrated at *B* is known as a "progressive" type. The go and not-go sections are combined in a single unit, and the minimum allowable diameter is followed by the not-go end. According to the American Gage Design Standard, both types are for diameters ranging from 0.059 to and including 1.510 inches.

The advantage of go/not-go or limit gages is that they permit the inspector to determine quickly whether or not a particular dimension of an article falls within the prescribed limits without the labor of adjusting or reading a measuring instrument. The limits of such a gage are, of course, those given by the tolerance.

While fixed limit gages simplify the work of the inspector, they are not desirable on precision processes where it is also important to

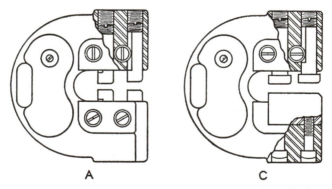

A   C

Fig. 8-4.   Two forms of American Gage Design Standard limit or "snap" gages. This form of gage shows if parts are within the maximum and minimum limiting dimensions.

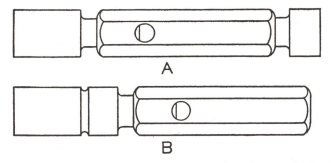

Fig. 8-5. Two forms of American Gage Design Standard limit gages for holes.

know *how far* or *to what degree* a product is inside or outside the tolerance. Knowledge of this degree of variation often furnishes important information on how and where equipment needs adjustment. An ordinary limit gage can only be used to indicate whether an article *does* or *does not* conform to specified tolerances. No attention is paid to the degree of deviation from the specification.

It is important to note that from the special use of limit gages just referred to, the term "go/not-go" or "go/no-go" testing has been adopted to cover all types of inspection that determine whether some product is or is not satisfactory, conforming or nonconforming, regardless of the degree of variation from the specifications. Examples of such general go/not-go tests can be found in the visual examination of steel bars for the presence or absence of sulphur stringers, scabs, pit and chafe marks, heavy seams, and bends; or the examination of polished parts to check whether or not they have the proper finish. Articles are either considered "satisfactory" or "defective" as a result of the examination. They either "go" or do "not-go."

Variables Tests

In contrast to go/not-go tests which merely indicate whether an article is satisfactory or defective, variables tests also indicate the *degree* of deviation. This may be shown either in inches, pounds, Rockwell units, or whatever other scale is applicable for the particular product and quality characteristic.

To illustrate, the ultimate shear strength of certain spot weldings is not supposed to be below 550 pounds per weld. Tests performed on a three-piece sample yield the following results:

Sample No.	Shear Strength
1	580 lbs.
2	560 lbs.
3	560 lbs.

The results show not only that the three welds are within the specification limit of 550 pounds, but also the *extent* to which each differs from it. Variables tests are used to determine many different types of quality characteristics, such as dimensions of machined product, hardness of heat-treated steel, percent moisture content, electrical resistance of wire, and many other items. The one factor common to all these tests is that quality is measured along a numerical scale such as inches, pounds, ohms, or any other unit of measurement. Examples have been given in previous chapters.

For simplicity, the results of variables tests are often converted to go/not-go data. The three spot welds previously mentioned might have been reported merely as satisfactory without regard to the numerical result.

In that case the information regarding actual shear strength in pounds would not be recorded. While this simplifies the work of the inspector, it is not desirable on processes where it is important to know *how far* a product is inside or outside specifications, so that necessary changes can be made accordingly and without loss of time. Thus, the above mentioned tests of shear strength, if properly recorded by the actual pounds, will indicate any tendency for successive samples to break at loads which are close to the allowable minimum of 550 pounds. Adjustments may therefore be made in the manufacturing process before any defective material testing below this value makes its appearance.

Implications for Process Capability

Go/not-go data indicate an attribute of a product, such as whether or not it conforms to certain specifications. If we were to state process capability on the basis of attributes data, we would be confined to noting the proportion of product that falls within and outside certain limits. This is an awkward approach, which moreover fails to portray the frequency distribution pattern that characterizes the capability of the production process.

If we are to employ attributes in describing process capability, it follows that the only type of control chart likely to be usable is a percent defective chart. While there is nothing wrong with such usage, it should be noted that such a chart usually requires from 10 to 20 times as much inspection and testing to acquire the reliability (that is, the same steepness of the operating characteristic curve) as the corresponding chart based on variables (specification-oriented or computed from center lines) would call for.

When variability has gone out of control, go/not-go data would indicate this by the fact that now off-standard units of product are found at both the high and low side of the tolerance. Usually, this approach is far less sensitive than the control chart for ranges.

In summary then, although go/not-go gaging has its place as a quick means of checking dimensional characteristics of product, for the purpose of evaluating process capability, establishing realistic tolerances, and maintaining effective control, variables testing is preferable.

Control Charts in Metrology

When testing involves complex or highly sensitive procedures and instruments, dual control charts—one on processing and the other on standard samples—may be needed. Carefully maintained units or reference specimens of practically unvarying characteristics comprise the standard sample. Three configurations may occur:

1. Both production and standard sample are (A) above upper or (B) below lower limits. Faulty testing is most likely to be responsible.
2. Processing is out-of-control while the standard sample is inside limits. Production is probably at fault.
3. Production is in control but the standard sample is not. This puzzling, slightly paradoxical situation may be produced by a variety of factors. For example, testing of the standard sample may have been faulty, but process testing was proper. Or else, the process may be off in one direction (say toward the upper limit) while testing is off in the other (toward the lower limit).

An illustration of configurations (1) and (3) appears in the example portrayed in Fig. 8-6. As always, when an off-standard condition occurs, checking and investigation must continue until the cause has been located and corrected.

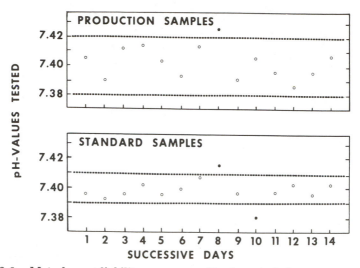

Fig. 8-6. Metrology reliability assurance. Dual control charts, one for the production process and the other for standard samples, are maintained.

Various refinements in the use of standard samples have been developed. For example, a variety of samples may be used, and at any one time the tester makes determinations "blindly," in the sense that he does not know the value expected. While some people may consider this approach needlessly time-consuming, in other situations it is viewed as an important aspect of quality assurance through reliable metrology.

REVIEW QUESTIONS

1. When one says: "Quality cannot be inspected into a product, it must be built in," what is really meant?
2. How do specifications and tolerances differ from control limits?
3. What is the definition of process capability?
4. Assume that process capability, in terms of grams per filled container, is given in terms of ±10. The product specification, however, allows tolerances of only ±8 grams. Among the actions open to the manufacturer are the following: (A) Decline to accept the order, (B) accept the order but screen out all nonconforming product, (C) try to improve his processing. What economic and technological factors might be important in arriving at a decision?
5. The manufacturer just discussed is able to improve process capability to ±9 grams. What, then, is his process standard deviation?

6. Assume that the manufacturer produces 1,000 units, given a process capability of ±9 grams and a tolerance requirement of ±8 grams. If the process average is well centered, what percentage of off-standard product may be expected? What would be the number of defectives?
7. Why is an economy and efficiency-minded producer keenly interested in investigating capability-tolerance interactions? How can both producer and consumer benefit from such analyses?
8. Using the nylon filament strengths from the prior Chapter, Review Question 9, what is the process capability?

In Appendix 3, Case 4 provides practice in the analysis of process capability. The data in Cases 2, 3, and 8 also provide an opportunity to determine process capability as useful management and engineering information.

CHAPTER 9

Vendor-Vendee Relations

Although we have previously referred to the vendor's application of control charts and sampling plans to assure the quality of his outgoing product and to the vendee's reliance on acceptance sampling to prevent substandard product from entering his plant, the relations between seller and buyer, supplier and user, producer and consumer are more complex and interlocking.

Interdependence

Time was when the purchasing of outside products was a relatively simple matter. The types of items involved were few, time pressures were usually not urgent, and widely recognized standards and specifications defined various quality grades. The twentieth century has changed all that, and the pace of change has accelerated with each decade. A vast array of new materials and new products has arisen. Assemblies are reaching demanding degrees of complexity at the same time as performance requirements are increasing and allowable tolerances are being narrowed. Multilevel subcontracting becomes necessary to obtain products engineered for specific uses. In order to assure proper coordination of all activities, vendor-vendee planning on all important fronts is an essential. This includes the design of the product, the development of processes, provision for inspection instrumentation, quality control and quality documentation, and interlocking scheduling of operations and delivery dates.

The reasons for this interdependence among vendors and vendees are manifold. The main contractor has committed himself to a quoted bid price, at a tight delivery schedule. Moreover, specifications and tolerances have been set at demanding levels. Any failure in his own operation will result in his inability either to produce within anticipated cost ranges, to meet the delivery schedule, or to submit product of requisite quality. Similarly, any failure at the subcontractor's plant will cause problems.

It is obviously too late to complain about poor quality of incoming shipments when the supplier (1) is not qualified to deliver to the specifications and tolerances required, (2) may not be fully aware of just what is implied by the specifications or tolerances, or (3) does not have a continuous quality control program of his own to assure the acceptability of his output. Any one of these failures can cause a chain reaction of devastating proportions. The problem becomes aggravated when a subcontractor in turn relies on outside components or raw materials.

Vendor as an In-House Department

The effect of these interdependencies is that, to minimize the risk of failures, the vendor must be looked on in a manner akin to an in-house department. Numerous meetings must be held to iron out technological problems, including provision for technological assistance and exchange visits to familiarize personnel with important production and product relationships. Prior to the appointment of a supplier, it is customary for inspection teams of the vendee to visit the plant. Among the things that the team looks for are the following:

1. Are the vendor's facilities adequate to provide the components needed in the quantities and at the productivity rates required?
2. What is the status of inspection and quality control? Are incoming materials checked? Are sampling plans and control charts applied at essential points during production? What assurance of outgoing quality is maintained? How often are gages checked and calibrated?
3. What are the qualifications, performance, and workmanship standards of the operators? How quality-conscious are they? What help do they receive in their work? What programs of quality motivation are in use, and how effective are they?
4. Are management and engineering qualified to turn out the quality and quantity of the particular product we need? Do the engineers understand the specific problems and requirements of the products to be produced for us? Is the attitude cooperative?

Most likely the vendor's plant will not fulfill all requirements. What is crucial, then, in the evaluation and eventual acceptance of a vendor is the *outlook for success* in a cooperative endeavor to upgrade management, engineering, and production to the levels needed.

Vendor's Quality Program

It is obvious that the vendor who has a quality control program in existence is at a decided advantage. The levels at which processes operate, the capabilities of equipment and the tolerances that can be maintained at various operating rates, the effectiveness of inspection and related types of information are readily revealed from existing records. From this base it is then a relatively straightforward matter to calculate the conditions, engineer the requisite changes, and make other adjustments as needed to produce to new tolerances for the vendee.

All of the tools discussed in the preceding chapters—such as acceptance sampling plans, process control charts, process capability studies, analyses of tolerance relationships, and the like—will be of value in providing essential information without loss of time. Moreover, an organization with quality control procedures in operation is more capable of adapting to new, tightened tolerance requirements than would otherwise be possible.

The documentation and certification of quality, particularly for characteristics that tend to be hidden in subsequent assembly, are especially important and can be performed effectively in only those organizations where control charts, sampling plans, and process capability studies are in active use on all major equipments.

The work of the vendor inspection team is minimized likewise when abundant information, based on the vendor's quality control records, is available. As a result, spot-checks may suffice where otherwise lengthy investigations might be needed.

Vendor Evaluation

Not all procurements involve such advanced technologies, stringent requirements, or otherwise crucial aspects as to call for continuing, intense exchanges of information, documentation, and visitations. Nevertheless, efficient operation of purchasing functions should be based on past performance of a vendor, on an evaluation of competitiveness of prices, punctuality in meeting delivery promises, and conformance to specifications and tolerances for individual units and production lots.

It is apparent that low price is meaningless without punctuality and quality. Similarly, high quality is of little value if the goods are not received on time to mesh with the purchaser's production and delivery schedules.

We shall consider, in the following, some of the ways in which vendor quality performance may be documented and evaluated.

Quality History Records

In Fig. 9-1 a quality history record for incoming materials is presented. The essential information includes the following:

1. Applicable time period.
2. Number of lots and units accepted during that period.
3. Number of lots and units rejected during that period.
4. Percent of defective product found during sampling inspection for acceptance of incoming shipments. This information is recorded in graphic form on the "quality chart" section of the recording form.

Assuming now that there are two or three suppliers, it should be a simple matter to compare quality performance among them. The example presented involves percent-defective inspection.

In Fig. 9-2 an illustration of variables sampling is presented. The calculation steps in columns 9 through 15 lead to the following:

1. The time period is given in 9.
2. Average range and grand average appear in columns 10 and 11.
3. The normal-distribution value of z is found by dividing the difference t of column 12 by the standard deviation in 13.
4. The standard deviation was obtained by multiplying the average range by the conversion factor. With sample sizes of 6 (see box 4 of the form), the conversion factor is 0.40 (rounded) as shown in box 5.
5. The value $t/\sigma = z$ is now entered in the normal curve table (Chapter 6, Table 6-1) to find the proportion of product falling outside the range of z. This is the percent of defective or "nonconforming" product. In other words, column (15) shows the proportion of product with strength below the minimum of 36 pounds specified.

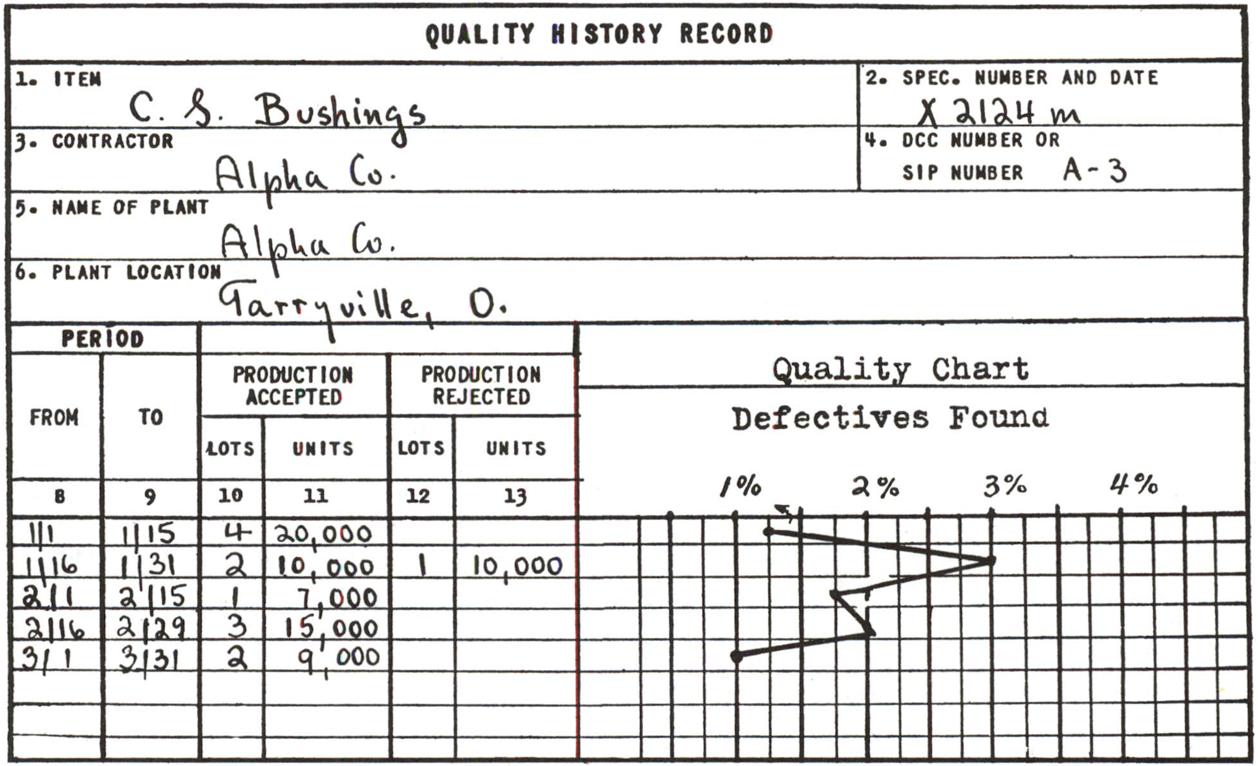

QUALITY HISTORY RECORD

1. ITEM C. S. Bushings

2. SPEC. NUMBER AND DATE X 2124 m

3. CONTRACTOR Alpha Co.

4. DCC NUMBER OR SIP NUMBER A-3

5. NAME OF PLANT Alpha Co.

6. PLANT LOCATION Tarryville, O.

PERIOD FROM	PERIOD TO	PRODUCTION ACCEPTED LOTS	PRODUCTION ACCEPTED UNITS	PRODUCTION REJECTED LOTS	PRODUCTION REJECTED UNITS
8	9	10	11	12	13
1/1	1/15	4	20,000		
1/16	1/31	2	10,000	1	10,000
2/1	2/15	1	7,000		
2/16	2/29	3	15,000		
3/1	3/31	2	9,000		

Quality Chart
Defectives Found
1%
2%
3%
4%

Fig. 9-1. Raw materials quality record, used by purchasing department as a guide in placing future orders.

QUALITY HISTORY RECORD

1. Product: Barrier material, flexible	2. Specification: Min. Strength = 36 kg.
3. Supplier	4. Sample Size: 6 / 5. Factor F_r: 0.39
6. Location	7. Prep. By / 8. Dates: From To

9. Period From	9. Period To	10. Average Range	11. Grand Average	12. t = (2) − (11), or (11) − (12)	13. σ = (5) x (10)	14. z = t/σ	15. Nonconforming, %
2/4	2/16	7.69	40.95	4.95	3.0	1.65	5.0
2/18	2/29	7.18	39.48	7.28	2.8	2.6	0.5
3/3	3/14	9.23	41.03	5.03	3.6	1.4	8.1
3/17	3/28	4.87	40.61	4.61	1.9	2.4	0.8
4/3	4/11	6.91	43.75	7.75	2.5	3.1	0.1
4/15	4/20	6.12	42.60	6.60	3.0	2.2	1.4

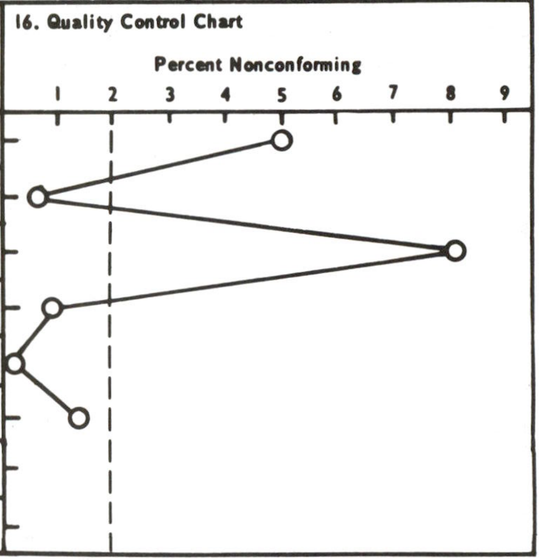
16. Quality Control Chart
Percent Nonconforming
1 2 3 4 5 6 7 8 9

Form reproduced by permission from P.O.M. Co., Industrial Consultants.

Fig. 9-2. Quality history record based on variables testing.

Again, a plot of past performance is provided (section 16 of the form), facilitating comparison among suppliers. Normality of the incoming shipments, at least to an approximate extent, is assumed. This assumption can readily be checked from accumulated data.

Acceptance Control Charts

Where certain items are purchased on a regular basis from a particular source, control charts may be established to serve as a basis of acceptance or rejection of lots. This activity may occur either at the vendor's plant or at the vendee's facilities. Assurance should be had, though, that shipments identify rational production lots, such as output from the same machine within a limited time period.

When the AQL is practically zero (or, more specifically, 0.1 percent), the types of control limits for specification-oriented charts (given in Chapter 4) are applicable. Often, however, a larger AQL is needed. Then, relabeling the control limit as a "rejection limit," we use the formula:

Rejection Limit $=$ Specification Limit $\pm$ (Average Range) $\times F_a$
where F_a is given in Table 9-1.

Table 9-1.   Factors for Acceptance Control Limits

Sample Size	Acceptable Quality Level (Percent Defective)						
	0.1	0.5	1.0	2.0	3.0	4.0	5.0
	Factor F_a						
2	1.49	1.03	.81	.57	.41	.30	.20
3	1.14	.84	.69	.53	.43	.35	.29
4	1.02	.77	.64	.51	.43	.36	.31
5	.94	.72	.62	.50	.42	.37	.32
6	.90	.69	.60	.49	.42	.37	.33
8	.84	.66	.57	.47	.41	.37	.33
10	.80	.63	.55	.46	.41	.36	.33
12	.77	.61	.54	.45	.40	.36	.33
15	.74	.59	.52	.44	.39	.36	.33
20	.71	.57	.50	.43	.38	.35	.32

NOTE: When AQL $=$ 0.1 percent or practically *nil*, $F_a = F_s$ (previously tabulated in Chapter 4). Other values of F_a were found by substituting in the formula for F_s the appropriate z-values in place of the value $z = 3.1$ for 0.1 percent of product outside the range of $3.1z$.

For example, the material purchased under a minimum strength specification of 36 pounds (see Fig. 9-2), was found to have an average range, $\overline{R}$, of approximately 7 pounds, based on sample sizes of 6 tests per lot. The agreed-upon AQL was 5 percent. We then proceed as follows:

1. From the tabulated values of F_a, find 0.33 under the AQL $= 5$ column in the row for sample size 6.
2. Multiplying $\overline{R}$ times F_a gives $7 \times 0.33 = 2.31$ pounds.
3. Add this 2.31 to the minimum specification of 36, obtaining 38.31 as the rejection limit.

No upper limit is needed. The higher the strength, the more acceptable is the product.

The acceptance control chart, in common with all charts on which averages are plotted, does not show up sudden increases in variability. Accordingly, it may be desirable to use a chart for ranges as well. Beyond this, continued close liaison between vendor and vendee is necessary to assure that lots inspected are also rational production lots. The importance of rational lots has been stressed at several occasions, as in Chapter 2.

The acceptance control chart is more technically known as a "control chart for variables, based on acceptable quality levels in percent defective." The term "variables" refers to the fact that measurements are made and used in computing limits. The limits are, however, established in a manner that allows an agreed-upon (small) percentage of off-standard product, the AQL. Additional techniques for acceptance control by variables, involving AQL's in percent defective, are provided in *Mil-Std. 414 Sampling Procedures* and *Tables for Inspection by Variables for Percent Defective*, U.S. Government Printing Office.

Frequency Distribution Patterns

Control chart results could be misleading when distributions are markedly non-normal. For this reason as well as for the sake of minimal use of statistical methodology and visual impact, there are many organizations that rely heavily on the presentation of frequency distribution patterns. Such an approach involves a good deal of testing, but often the costs involved are not prohibitive.

Figure 9-3, for example, is a quality documentation report, sub-

QUALITY REPORT No. 4823

PRODUCT _COMPRESSION SPRING_ CUSTOMER ___ QUOT. No. _87884-8_
CHARACTERISTIC _P @ L OF 0.180"_ PART NO. _9008243_ REF.
INSP. METHOD _217-PM_ SPECIFIED LIMITS _200 GRAMS ± 5% @ 0.180"_
SAMPLE DRAWN _AFTER CAD. PLATE_ EQUIV. INSP. LIMITS _190 - 210 GRAMS @ 0.180"_

HPS ORDER No.		80710 (15M)	82444 (5M)	
CUSTOMER ORDER No.		116649	117243	
INSPECTED BY		RHD	E.B.C.	V.G.S.
DATE		1-6-48	1/7/48	2-9-48
FORCE (P)		LOT 1	LOT 2	
	MIN.			
190				
2		‖	‖	‖
4		ℍ Ⅲ	ℍ Ⅱ	ℍ Ⅲ
6		ℍ ℍ ℍ Ⅱ	ℍ ℍ Ⅲ	ℍ ℍ ℍ Ⅲ
8		ℍ ℍ ℍ Ⅰ	ℍ ℍ ℍ Ⅲ	ℍ ℍ ℍ ℍ Ⅲ
200		ℍ ℍ Ⅰ	ℍ Ⅲ	ℍ ℍ ℍ ℍ ℍ
2		Ⅲ	ℍ	ℍ ℍ ℍ Ⅱ
4			Ⅰ	ℍ ℍ
6				ℍ Ⅰ
8				‖
210	MAX.			Ⅰ
		OK	OK	OK
		8100 PCS.	7M PCS	6100 PCS
		C.J.M.	C.J.M.	C.J.M.

HUNTER PRESSED STEEL CO.
FORM 561-10-15.17-47

Fig. 9-3. Vendor's quality report. Frequency distributions of successive shipping lots of compression springs are shown.

mitted by the vendor, and certifying the results of his own check of the quality of outgoing product. If occasional spot-checks by the purchaser confirm the supplier's quality reports, then the latter are generally accepted on face value.

Typical frequency distributions encountered in the inspection of incoming shipments are noted in Fig. 9-4. A wealth of pertinent information is revealed, and much can be gleaned from the pattern of the distribution, in relation to upper and lower tolerance limits. This information refers to both the quality of the lot as a whole and the likely causes of off-standard performance.

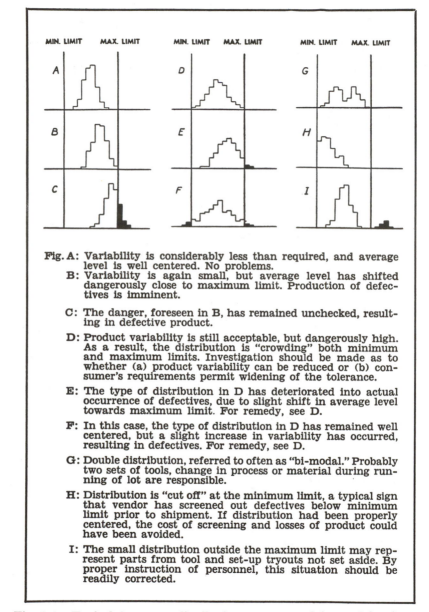

Fig. A: Variability is considerably less than required, and average level is well centered. No problems.

B: Variability is again small, but average level has shifted dangerously close to maximum limit. Production of defectives is imminent.

C: The danger, foreseen in B, has remained unchecked, resulting in defective product.

D: Product variability is still acceptable, but dangerously high. As a result, the distribution is "crowding" both minimum and maximum limits. Investigation should be made as to whether (a) product variability can be reduced or (b) consumer's requirements permit widening of the tolerance.

E: The type of distribution in D has deteriorated into actual occurrence of defectives, due to slight shift in average level towards maximum limit. For remedy, see D.

F: In this case, the type of distribution in D has remained well centered, but a slight increase in variability has occurred, resulting in defectives. For remedy, see D.

G: Double distribution, referred to often as "bi-modal." Probably two sets of tools, change in process or material during running of lot are responsible.

H: Distribution is "cut off" at the minimum limit, a typical sign that vendor has screened out defectives below minimum limit prior to shipment. If distribution had been properly centered, the cost of screening and losses of product could have been avoided.

I: The small distribution outside the maximum limit may represent parts from tool and set-up tryouts not set aside. By proper instruction of personnel, this situation should be readily corrected.

Fig. 9-4. Typical frequency distributions encountered in receiving inspection. Lot patterns and their position relative to the tolerances provide valuable clues to vendor quality and likely causes of off-standard performance.

Summary

Several approaches for control of quality in vendor-vendee dealings have been presented. In addition, general philosophies and underlying reasons have been discussed. Which specific approach and what particular tools should be used in a given instance? It is apparent that such factors as quality requirements, testing costs, and confidence in the vendor's capability must be weighed in arriving at a sound decision.

REVIEW QUESTIONS

1. What are the principal reasons for interdependence among vendors and vendees, such as a main contractor and a subcontractor?
2. In what way may a vendor be considered as an "in-house" department?
3. Why is vendor evaluation necessary before a contract for essential components is signed?
4. Quality history records among three vendors show them to average the following percentages of defectives on certain similar electronic components: 1, 1.3, 1.8. Despite this fact, the purchasing department consistently offers the last-given vendor (1.8 percent defectives) the largest contracts. What considerations would justify such action?
5. What is the purpose of acceptance control charts?
6. Under what conditions would you feel inclined to accept the producer's certified frequency distributions submitted with each shipment?
7. What quality control tools are available for checking that in general a producer's submittal of product frequency distributions is accurate?
8. Required strength for a pressure relief valve is 10 ± 2 kg. per square centimeter. A supplier's quality history, based on the testing of 10 valves per shipment, provides the results below for the year's four quarters. Approximately 10 shipments are received per quarter.

	First	Second	Third	Fourth
a. Grand average	10.2	9.0	11.0	10.5
b. Average range, $\overline{R}$	2.2	1.0	1.5	2.0
c. Estimated standard deviation, σ, applying factor $F_r = 0.32$ from Table 6-2 to $\overline{R}$	0.7	0.32	0.48	
d. Difference between grand average and nearest tolerance limit	1.8	1.0		
e. $z = \text{Difference}/\sigma$	2.6	3.1		
f. Estimated percent nonconforming valves (from Table 6-1)	0.5	0.1		

What are the values of the missing entries in this schedule?

CHAPTER 10

Organizational Aspects

The development of modern and more effective methods of inspection and quality control has brought into the limelight the problem of fitting these activities into the organization in a manner that will assure the greatest effectiveness.

Organizational Patterns

Modern quality control has provided production management with a new scientific tool that has shown itself capable of indicating impending trouble even before defective product makes its appearance. To make most effective use of this tool, three different types of organizational approach have been used. Each of them has its distinct advantages and disadvantages, making one or the other preferable for a particular type of manufacturing plant.

The three most commonly used organizational patterns for quality control are presented in diagram form in Fig. 10-1. They are based upon the fundamental fact that quality of product depends upon the coordinated activities of the production, engineering, and inspection departments.

Advisory Organization

Under this type of organization (see diagram at left, Fig. 10-1) inspectors, inspection supervisors, and other inspection specialists are established as "process advisors" to production and engineering. They exercise no authority.

Advantages

1. No great changes are required. Possibility of disagreements between inspection and production is minimized.

2. If management is progressive, quality control advisors will be able to obtain cooperation in establishing quality control, particularly if how it reduces costs can be shown.

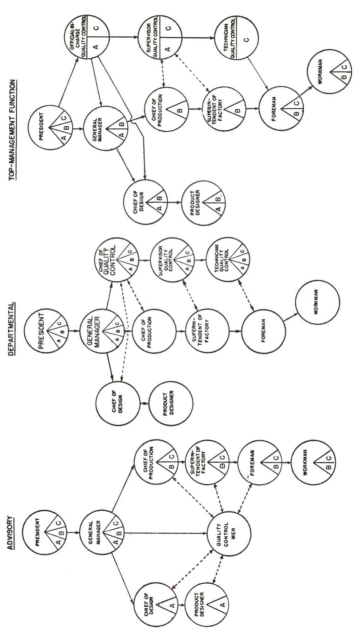

Fig. 10-1. Organizational patterns for quality control. Full or solid leader lines on diagram indicate flow of authority and responsibility. Dotted lines mean "confer and advise." *A* indicates responsibility for quality standards, *B* responsibility for quality of product, and *C* responsibility for determining and reporting quality of product.

Disadvantages

1. Lack of authority. Production executives can heed or ignore good advice, and "top" management may not be informed about the success of quality control in the plant.

2. If plant operation follows "traditional methods," human inertia will prove a stumbling block to good results.

3. Quality control advisor has no definite responsibility; therefore he lacks incentive to produce results.

Departmentalized Organization

Under this type of organization, definite duties and authorities are established in the form of an inspection and quality control department. (See central diagram, Fig. 10-1, for details of responsibilities.)

Advantages

Quality control chief has definite responsibility; therefore he has an incentive to produce results accordingly.

Disadvantages

1. Quality control chief often tends to build up his prestige at the expense of production and engineering departments. May create negative reaction to the effect that quality is no concern of production and engineering executives, and that it is entirely the responsibility of the quality control department.

2. Quality control chief cannot direct that corrective action be taken.

3. Because quality cannot be *inspected* into a product but must be designed and built into it, departmentalized quality control is likely to be reduced to the status of a "traffic cop" separating good production from bad. Thus the "quality control" department may become merely a department for screening good product from bad.

Top Management Function

Making quality control a functional part of top management means placing the quality control chief (an officer of the company, if desired) in a position similar to that of controller. Specific duties may be assigned as shown by the diagram.

Advantages

1. The quality control chief can place responsibility for quality where it belongs, direct that corrective action be taken where necessary, and establish company policy as regards quality.

2. If functions have been carefully defined, this gives the quality control personnel positive duties and responsibilities as well as a chance for adequate reward.

3. Authority is established to accumulate facts and figures and to direct what is to be done about them.

Disadvantages

1. Functional organization may produce resentment on the part of those who believe that their authority is impaired.

2. Danger of improper distinction of duties.

A conclusion regarding the overall merit of any one of the methods just outlined cannot be stated, since much depends upon the individuality of a particular business organization and the type of manufacturing. In chemical manufacturing, for example, where considerable attention is paid to tests and test results, the advisory type of organization may be the best. Such an organization, however, usually would be ineffective in many types of precision metal work. Here the departmental organization may be preferable in small plants and functional organization in large plants. The personal element involved—such as the attitude of management, the extent to which management is ready to promote quality control, and the personality of the quality control chief—plays a decisive role in the choice of a successful system.

Obtaining Cooperation and Compliance

To achieve effective quality control, it is of paramount importance to obtain the cooperation of production people and to induce them to comply readily with whatever necessary corrections are brought out through inspection. This in turn requires not only proper organization and management, but fair and impartial inspection, as well as an understanding of the aims of inspection by production personnel.

In order to present clearly the important principles of quality control and its effective use, simple illustrations of principles will be a helpful means of supplementing an education and training pro-

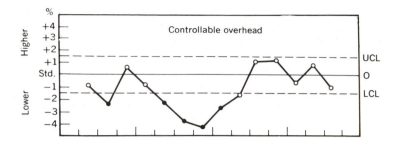

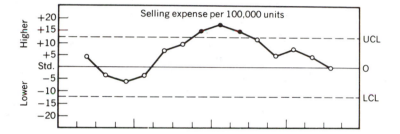

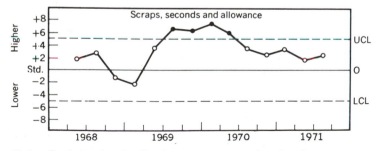

Fig. 10-2. Organizational effectiveness, as revealed by financial and administrative control charts. Performance is measured as deviation from valid, established standards.

SOURCE: N. L. Enrick, "The Control Chart as a Management Tool," in *Quality Assurance*, vol. 8, no. 7, p. 38–40.

gram throughout the plant. They may be incorporated in standard practice manuals or be exhibited as posters in connection with whatever methods are in use at the particular plant involved.

Organizational Effectiveness

Quality control applications can be readily made to become measures of organizational effectiveness. Three illustrations, taken from a wealth of modern uses, are presented in Fig. 10-2. The following purposes are served:

Controllable Overhead

Financial management as well as the plant manager and his staff will be principally concerned. Controllable overhead represents such items as maintenance, inspection, quality assurance, and warehouse operations over which management has at least some degree of control. In the example, shortages of qualified help and slow management reaction resulted in inadequate provision for preventive maintenance and quality inspection. Charts for such factors as property taxes and depreciation, which cannot be changed in the short run, are omitted as "fixed overhead."

Selling Expense

Of principal interest to marketing and sales, this chart reveals the cost effort required to sell the plant's output. The plant manager should also be interested, since product design and quality have an important bearing on salability. Top management may use the chart to resolve questions of promotional allocation, marketing channels, effectiveness of product development, production, advertising, and production-inventory-sales coordination. To a large extent, the period of relatively high selling expense portrayed on the chart represented the effects of quality inadequacies (see prior chart).

Scrap, Seconds, and Allowances

This information is vital from both a financial aspect, the cost of poor quality, as well as from an administrative consideration of whether or not quality control is "doing its job" or receiving proper support. We note a period of high costs, undoubtedly caused by the previously observed months of inadequate inspection and maintenance in production.

Typical additional charts are tabulated below. Further, comparisons among production plants, warehouses, and branch offices are feasible.

Summary of Typical Administrative-Financial Charts

Category	Comparative Breakdowns
Productive labor	Departments
Indirect labor	Departments
Clerical salaries	Administration, sales, bookkeeping, computer
Raw materials	Actual *versus* standard consumptions
Repair costs	Departments
Supplies	Types (such as cutting oils, dyes, finishing compounds)
Power consumption	Electricity, gas, coal, etc., by uses
Quality	Rework, "invisible" losses or shrinkage, scrap sold *versus* scrap recorded by departments, returns and allowances by products
Turnover rates	Departments
Machine efficiency	Downtime by cause, running time percentages by departments, overtime hours by cause
Receiving delays	Average time from receipt through inspection, by types of incoming products
Inventories	Raw stock, materials, goods in process, semifinished and finished items
Marketing	Gain rate or loss rate of old customers, by product groups
Diversification	Number of models, styles, colors offered by product groups

In all instances the management-by-exception principle should be the rule: So long as actual performance is within allowable limits and no adverse trends are detected, no action is taken. It is only when these conditions are violated—the exception situation—that management takes action with the investigation of causes and a search for corrective measures.

Planning for the future, goal setting, and efforts to enhance financial status, administrative performance, and organizational effectiveness are readily developed on the basis of information on past averages and patterns of variation, as revealed by the control charts.

A methodology that, in many a plant, starts as an approach to control and enhance an area of operations (production), is thus extended to serve as an aid in improving overall organizational performance. Where this work has been done carefully and properly, the benefits have been considerable.

Total Quality Control

Planned and goal-conscious work is essential in the installation, maintenance, and growth of economical and effective quality control and product reliability. Otherwise, organizational and managerial efforts may well be frittered away on conflicting, fragmented, and uncoordinated activities.

An orderly cycle, such as portrayed in Fig 10-3, constitutes the core of what is generally known as "total quality control," or "total quality-reliability programs." The term "total" refers to the closing of all gaps in the control loop. Not only is there testing and inspection for quality and reliability, from raw stock through various processing stages until finished product, but feedback and planning pervades the entire organizational setup. In particular, consumer and market acceptance of the product, customer complaints and field-service reports serve for the purpose of review and redesign of components and complete equipment. Processing operations, as well as quality and reliability testing and inspection, are planned in accordance with experience in production and product usage.

Close attention is paid to the factors of design, workmanship, and

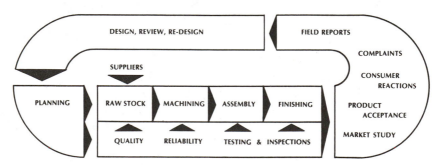

Fig. 10-3. Total quality-reliability program. A continuing, planned cycle of quality, reliability, and cost improvements is involved in the design and redesign of products, processes, inspection, and testing.

uniformity in relation to (1) the quality and reliability problems of products and production, (2) the tools of quality and reliability attainment and (3) the goals of the overall program. The aims encompass cost and marketability aspects, from both a short-range and long-term viewpoint, as noted in Fig. 10-4.

Systems and Procedures

Since quality assurance work cuts across organizational lines, it is desirable that systems and procedures for quality control work be established carefully, based on exploration of pertinent factors and implications with all parties concerned. Procedures are usually laid down in writing, subject to occasional reviews and revisions.

An approach that is useful in connection with this work involves application of so-called *decision matrices*. A simplified illustration, involving the problem of what to do when an incoming lot has been rejected by receiving inspection, is presented in Fig. 10-5. The upper section represents problem conditions and configurations, while the lower section presents the applicable decisions. For configuration 3, for example, the following is the standard procedure:

IF defects are (a) serious and (b) can be repaired, but (c) no repair labor is available, and (e) the plant is not in urgent need of the shipment . . .

DESIGN	WORKMANSHIP	UNIFORMITY
The Quality and Reliability Problem		
How useful, serviceable and reliable is the design? What costs will be involved? How attractive and appealing will the end product be?	Does a sense of quality-mindedness pervade the entire organization, executives, staff, supervisors and workers?	Is product variability kept within allowable tolerances? Do these tolerance limits balance cost and quality-reliability needs?
Tools of Quality and Reliability Attainment		
Experience. Review of records of past performance, failures, complaints. Field tests, life tests, pilot studies and surveys. Market analysis and research. Revision of designs based on these types of data. Value analysis. Value engineering. Automation.	Creation of quality awareness of employers, supervisors and staff. Employee selection and training. Quality incentives. Process and product standardization. Quick corrective action where needed. Effective preventive maintenance.	Analysis and reduction of process and product variability. Process and product uniformity research. Use of precision instrumentation and gauges. Statistical sampling plans and control charts from incoming materials to finished product.
Goals of a Quality-Reliability Program		
Functional, reliable and attractive products, with a realistic quality/cost ratio, meeting a market need.	Reputable quality, assured product reliability. Avoidance of defects and hidden faults and deficiencies.	A uniform product of precision manufacture, produced at reasonable and justifiable price.

Fig. 10-4. Role of design, workmanship, and uniformity in a total quality control organization.

PROBLEM CONDITION	PROBLEM CONFIGURATIONS (No. 1 TO 14)													
	1	2	3	4	5	6	7	8	9	10	11	12	13	14
a. Are defects serious?	O	O	O	O	O	O	O	O	●	●	●	●	●	●
b. Can they be repaired?	O	O	O	O	O	O	●	●	O	O	O	O	●	●
c. Is repair labor available?	O	O	●	●	O	O			O	O	●	●		
d. Does vendor agree to pay for repairs?	O	O			●	●			●	●				
e. Does plant need the product urgently?	O	●	O	●	O	●	O	●	O	●	O	●	O	●
DECISION CATEGORIES	**DECISION RULES**													
f. Return shipment.			√	√		√	√	√		√		√		√
g. Use product but seek allowance from vendor											√		√	
h. Repair product and use it.	√	√			√				√					
i. Seek urgent replacement elsewhere.			√				√							
j. Go to further tables for more instructions.			I		II		I		II					

Code : O = "yes," ● = "no," √ = "make this decision."

Fig. 10-5. Decision matrix. Answers "yes" and "no" under the various problem configuration columns result in a distinct problem pattern, for which response patterns are given under the decision-rules columns. For configuration 8, for example: "IF defects (in an inspected incoming lot) are serious, but cannot be repaired and the plant does not need the shipment, THEN apply decision *f*, which calls for a return of the shipment."

SOURCE: N. L. Enrick, "Tabular Logic in the Decision Process," in *Industrial Canada*, vol. 69, no. 6 (October, 1968) p. 32–35.

THEN proceed as follows: (f) return shipment and (i) seek urgent replacement elsewhere, then (j) go to another table (Table I, not shown as part of the presentation) for additional instructions.

It is not claimed that this type of action is generally desirable. All we are doing here is to illustrate the manner in which a decision matrix works. It considers all relevant problem conditions and their interacting configurations, from which procedures for action are developed. Each organization will work out its own procedures. Going beyond quality control, procedures for routing of product, operation of processes and other production, administrative and operational routines, can likewise be encompassed in decision matrix forms.

REVIEW QUESTIONS

1. What are the principal advantages of the advisory type of quality control organization?
2. In what way is a departmentalized organization (A) stronger or (B) weaker than an advisory set-up?
3. What dangers are inherent in setting up quality control as a top management function?
4. Why is it important to make efforts in every way possible to obtain cooperation of production people in the performance of quality control work?
5. It has been noted that the type of manufacturing and processing in a production organization has a bearing on the type of organizational setup most suited for the quality control program to work effectively. Can you discuss this further?

CHAPTER 11

Installing Quality Control

In concluding Part I of this book, we present the salient aspects involved in the installation of quality control procedures in a manufacturing plant.

The Pilot Run

One should not attempt to introduce the program all over the plant simultaneously. Instead, select one spot in production which appears most promising for success, and let it be the starting point. Only after this pilot application is operating properly will it be wise to extend installations to other machines and operations, production lines and departments, and finally to the entire plant.

The proof of each installation lies in the results attained. To make sure that an individual application will be workable under practical operating conditions, trial runs should be carried out first. Occasionally, such studies will show that tolerances, control limits, the AQL, or processing factors need adjustment. Only when we are satisfied about the feasibility of an application should it go into routine, regular usage.

Starting Points

When surveying the factory for quality control possibilities, one might look for the following points as most promising:

1. Wherever screening is in use, the possibility of substituting more economical sampling plans should be well worth investigating.
2. Trouble spots, involving excessive output of defectives, might be examined. Careful analyses of process capabilities and the application of control charts may be of considerable value in identifying causes of trouble and attaining improvements.
3. Instances of superficial inspection should be scrutinized. Often some unsuspected operations are producers of sizable quanti-

ties of defectives. They may be arrested by installing sampling plans.

Types of Inspection in Different Areas

The installation of quality control should be done with the needs of different areas in mind.

1. *Incoming materials.* Many organizations still maintain inadequate staffs to inspect incoming materials properly. Perfunctory examinations are not adequate in protecting the plant from the use of deficient products. Here a study is indicated to determine through lot-by-lot sampling whether the supplies, components, subassemblies, and other materials going into production conform to specifications and tolerances. In other plants, where screening has been applied in the past, it may be found that lot-by-lot sampling is not only more economical but also more effective in spotting bad lots.

2. *Partly Finished Product.* Inspection here involves the checking of product while it goes through the various productive phases. Since on certain operations defective work will be difficult to avoid, a control chart or sampling plan will be of service.

3. *End Product.* When end-product or final goods are inspected, screening is found to be the most widely prevalent method. As a rule, management will be reluctant to replace it with any other procedure. If all preceding production has been brought under good control, however, it is unlikely that more than a token quantity of unsatisfactory product is "slipping through." Screening may then become superfluous, to be replaced by sampling plans. On the other hand, where quality levels are such as to make screening a necessity, it should be realized that 100-percent inspection is in effect a routine operation, akin to any production operation. The efficiency, proficiency, and performance quality of such screening should be checked by introducing lot-by-lot sampling as a control.

Control Charts Versus Sampling Plans

In order to control the quality of work in process, we might either install control charts at the machine or inspect the product by means of sampling plans. The question arises: Which method is preferable? To simplify the discussion, let us assume that the predominant use of control charts is accompanied by variables measurements (scalar values, such as ounces of weight, inches of length, or degrees of temperature), while the sampling plans (in Chapter 2) involve go/not-go types of "satisfactory" or "defective" determinations on each unit of product.

Many processes can be controlled successfully by either method. As a matter of fact, different plants use different methods on the same product with equal results. For example: A certain shaft is required to be machined to a diameter of 3 inches + 0.0000 − 0.0015 inch. One plant uses a control chart, measuring each sample of four shafts and plotting the average on the chart. The other plant uses a sampling plan, checking each shaft with a go/not-go gage, and rejecting any lot where the sample indicates it contains too many defectives. Both plants are achieving satisfactory results; but this does not apply to all processes. In each individual case a decision has to be made on the basis of the following facts:

1. The information obtained from control charts is superior to that obtained from sampling plans, because control charts often warn when a process is going out of control even before defective product occurs. A sampling plan can only tell you that the process is out of control after the defectives have already been produced.

2. The sample size is much smaller under the control chart system, but sometimes each sample has to be tested very carefully in order to obtain a precise measurement or pointer reading. Sampling plans, although they require larger sample sizes, involve merely a simple test or visual examination to determine whether each item is satisfactory or defective.

3. The amount of figure work to be done is, of course, greater when the control chart is used, since we have to compute sample averages and process variability.

The advantages versus disadvantages are as follows:

	Advantages	Disadvantages
Control Charts	superior information small samples	more figure work more painstaking tests
Sampling Plans	less figure work simple tests	less information bigger samples

Report to Mr. S. Smith, Plant Manager

on

RECOMMENDATIONS FOR ESTABLISHMENT OF

MODERN QUALITY CONTROL IN THE PLANT

This report contains recommendations for the introduction of statistical quality control techniques in our production operations. These methods have been finding increasing adoption throughout industry as a means of economic and efficient control of quality.

In general, the techniques proposed for adoption are adapted from those recommended by the American Society for Testing Materials, the American Standards Associations, the American Society for Quality Control, various governmental agencies such as the Defense Department, and by many prominent industrial organizations. These methods are known as "statistical quality control." The word "statistical" here refers to the use of statistical methods and probability in calculations analyzing inspection data and in controlling production and product quality.

If the methods proposed here meet with approval, authorization is desired for carrying them into effect.

The Present System

In the light of modern concepts and recent investigations, it may be said that while our inspection system is technically correct, it has certain weaknesses in methods. These are:

1. Undue reliance is placed upon our few roving process inspectors to control production. Since they cannot be in all places at one time, process inspection is far from adequate and should not be relied upon as a primary means of quality control.

2. Undue reliance is placed upon 100 per cent inspection by bench inspectors to screen and sort out defective product. Spot checks have shown that from 8 to 12 per cent of the defectives in a lot have remained after screening. This observation merely bears out studies made in many industrial plants which showed that as a result of fatigue and monotony in screening inspection, as much as 15 per cent of all bad products may be overlooked. Thus, if a lot should happen to contain 8 per cent defectives, about 1 per cent defectives may still remain in it after screening has been performed.

The Solution of Present Difficulties

The existing quality problem will not be solved by hiring more roving process inspectors or by doing more bench inspecting. Neither will there be permanent results by instructing inspectors to be more watchful, since the majority of them are known to be doing as well as can be expected. Instead, it is recommended that an entirely different system of inspection be adopted: statistical quality control. This technique aims to utilize inspection data as much as possible in anticipating and preventing the production of defectives.

How Statistical Quality Control Works

The application of statistical quality control is best illustrated by showing how it would work in one department, for example, our Drop-Hammer Department:

Step 1: An inspector draws periodically a random sample from the blanks produced on each drop-hammer.

Step 2: When he has drawn the required number of sample pieces, he takes them to his inspection and gaging bench and gives them a very careful examination. If necessary, he will compare any questionable piece against standard samples showing borderline quality margins for surface imperfections, or check it against his Standard Inspection list showing what types of defects are classed as "major" and "minor."

Step 3: After the inspector has examined every piece in his sample, he counts the number of defectives found (if any) and compares them against a sampling table worked out for him by the Quality Control Department. This table tells him when to accept lots and when to reject them for poor quality, on the basis of previously established standards.

Step 4: If the total number of defective pieces found is equal to or smaller than the permissible number given in the sampling table, the inspector knows that in all probability the lot represented contains such a small percentage of defectives (if any) that it is acceptable. Consequently, the lot is passed to the next stage of production. But if the total number of defectives found exceeds the permissible number, this constitutes strong evidence that the lot contains an unacceptable percentage of defectives. The inspector must reject the lot, notifying the foreman so that corrective action on the process can be taken immediately to stop defective output. The bad lot itself is "detailed" to screen out all defectives by 100 per cent inspection.

3

Defects Prevention

The important difference between screening and statistical sampling is now apparent. Screening serves merely to separate defectives from good product. Statistical quality control is designed primarily to obtain inspection data for use either in preventing defective production, or in stopping it where it occurs, as quickly as possible. This is the principal problem in our plant.

Basis of Sampling Tables

Management must allow for the fact that in high speed production a certain small percentage of defective output is unavoidable. Known technically as the "Acceptable Quality Level in Per Cent Defective," this allowable proportion is determined from the following factors:

1. The cost of further processing defective pieces which will finally be scrapped.

2. The quality that may reasonably be expected under mass production conditions in each processing stage and operation.

3. The normal effect of the many variables of production on the uniformity of the process.

The sampling tables furnished to the inspector by the Quality Control Department take into account the Acceptable Quality Level (AQL), as well as the ever-present sampling risks of:

1. Erroneously accepting a bad lot.

2. Erroneously rejecting a good lot as "bad."

These risks can be reduced to a minimum by furnishing the inspector with sampling tables that have been determined from probability calculations and are supported by practical judgment.

Quality Improvement

It was pointed out that control is designed to detect and stop defective output when and where it occurs. Moreover, data accumulated over a period of time will indicate where additional training of operators is needed, where special maintenance of equipment is required, and where basic changes in methods of operation, equipment, or design may be desirable.

Primary Responsibility for Quality

The purpose of the modernized inspection department will be to help the production floor more effectively in turning out an

4

acceptable product. It does not replace the primary responsibility for quality, which rests with the foremen and workmen, as it has in the past.

Organization for Quality Control

In order to obtain effective quality control, it is necessary to establish a managerial relationship which will assure prompt action on the process whenever inspection data indicate lack of control. An important feature of a good program is that quality control is made a function of top management concern, and that the department responsible for quality control is independent of either Production or Engineering. Quality is the result of efforts by both Engineering and Production, and quality control should be free to establish the facts as they are.

H. Brown
Chief Inspector

Fig. 11-1.   Recommendations for installation of a quality control program.

Report to Mr. S. Smith, Plant Manager

on

REVIEW OF THE FIRST TWO MONTHS' RESULTS

OF OUR MODERNIZED QUALITY CONTROL PROGRAM

Balance Between Quality and Quantity

There are many problems that have arisen in our program of installing and maintaining an up-to-date quality control system, but they all reduce ultimately to one major problem: how to maintain a standard level of quality and at the same time obtain the quantity of output required. All other problems are merely contributing factors to this major one.

It became apparent soon after we installed the quality control program that attainment and maintenance of quality of production would be complicated by a multitude of factors arising from the need for quantity production. We found that we would have to travel a middle road between quality and quantity: guarding against excessive relaxation of quality standards to assure volume of production, while at the same time isolating unsatisfactory operations, operators, or machinery and equipment which could be corrected without appreciable loss in production.

This dilemma was aggravated by the fact that not enough experienced inspectors have been available to provide adequate supervision of operations plus skilled examination of product. For although inspection has always consumed a great number of man-hours at our plant (about 500 hours per week, as shown by a survey), the majority of qualified inspectors were found to be part-time production workers (die-setters, assistant foremen, piece-rate men) who were little interested in a full-time inspection. Personnel doing inspection 100 per cent of the time were found to be experienced in only certain types of operations (such as examination of tinning and pointing only, or grinding only, or plating only, etc.) and stationed at points where inspection was merely segregating but not corrective. For example, inspectors were sorting out bad blanks and die-cast bolsters from good ones before mounting in the Assembly Department, while they should have been working preventively in the departments where the defects actually occur. The extensive shifts in inspection to the most effective locations involved considerable un-learning and re-training, both in the technique of examining individual items as well as determining the acceptability of lots through sampling inspection.

1

Throughout the program, emphasis has been placed on installing lot-by-lot inspection and control charts at the points in production where they would do the most good in detecting incipient defective work and stopping the outflow of unsatisfactory products. As a result, the former high cost of 100 per cent inspection of articles prior to assembly has been eliminated.

Standardization of Inspection

Specifications for methods, quality of finish and workmanship are given for each item by the time-study and methods standard. Hence, it was necessary that the basis for accepting and rejecting lots be standardized for all operations so that every operator receives the same treatment. This task involved standardizing and unifying the methods, practices, and controls of all phases of inspection.

Standardization of inspection has been approached from two points of view: inspection of the individual item (technical) and inspection of the lot (statistical). The technical side of the problem is being solved by preparation of written Standard Inspection Lists and standardized product nomenclatures for all products and operations. The statistical side of the problem is being solved by collecting and analyzing inspection and cost data to arrive at proper values of the Acceptable Quality Level for each item and the issuance of the standard sampling plans based on these factors.

Gages and Templets

With the aid of Engineering, continued progress has been made in the development of gages and templets and automated instruments to control such important variables as contours of blanks after drop-hammer operations, the width of tangs before and after slush-casting, the thickness of blades after grinding and glazing, and numerous other quality items. The use of gages and automated equipment eliminated many difficult and time-consuming manual measurements.

Sampling Plans

Our type of manufacturing is at such high rate of speed and the value of the individual article produced is so small, relatively, that it does not warrant screening except in the Packing Department, where 100 per cent examination is done to prevent shipment of defective. It is necessary, however, to determine the quality of each lot in process following certain operations. After striking off blanks, for example, lot-by-lot sampling inspection is performed to isolate any drop hammers on which excessive scrap is being produced caused by wear, shifting of dies, or careless operation. This makes it possible to take immediate corrective action.

2

3

But that is not the only function of quality control sampling. In Finishing, for example, it must be ascertained that blanks are properly polished before plating. At the same time, it would not be economical to examine every tote box of 500 pieces, for many of them will contain only a few poorly polished pieces. Here again application of sampling plans solves the problem: if the sample indicates that the lot probably contains more defectives than permissible, the entire lot is detailed to 100 per cent screening inspection. Thus, the defectives are sorted out, to be reworked or scrapped. Preventive action insures a minimum of defectives after plating. The costly prospect of rework subsequent to plating is thus minimized.

Quality Improvement

Observations in many departments already show gratifying results as regards quality of product. This is borne out by the attached graphs representing departments where quality control inspection methods have been in effect for some time.

There has been a considerable decrease in scrap following installation of lot-by-lot inspection in Drop Hammer Department. Part of this decrease is due to psychological factors: the knowledge that his work is being systematically inspected makes the operator more quality minded. Previously nobody had seemed to care much. More important, however, is the fact that lot-by-lot inspection data were analyzed to single out worn dies, poor set-ups and, occasionally, carelessness.

A reduction of scrap produced in this department, from a yearly average of 2.5 per cent to approximately 1 per cent, also results in considerable weekly saving of costly raw materials. Detailed figures for stainless steel, nickel-silver, and brass were prepared by the Cost Department.

Reduction in cost of rework operations, such as to remove die marks and flashes, and chargeable to the Pressworking Department has decreased since installation of control charts in that Department. With the fast rate of output on our presses, it is easy for die marks to go unnoticed for some time. Yet, it takes a costly and strenuous polishing lathe operation to remove such marks.

Thus the Pressworking department, long recognized as a major source of quality headaches for both operators and foremen, has shown such considerable improvement that it was possible to eliminate screening of stainless steel blanks in the Assembly Department. The foreman of that department agreed that the percentage of defective blanks received there had declined to an insignificant level. Two other points where 100 per cent inspection has been performed in the past are scheduled to be replaced by lot-by-lot sampling also.

4

Striking improvements in quality were also achieved in Plating and Finishing.

Cost Savings

While the forthcoming quarterly operating statement and cost-benefit analysis prepared by Accounting reflects considerable tangible savings due to a reduction of scrap and rework, the more significant long range gains arising from our quality control program cannot be stated so readily.

The most important factor here probably is the increased good-will among foremen and workmen created through orderly and systematic inspection in which decisions to reject lots or stop faulty operations and set-ups are not based on "arbitrary" judgment, but on the impartial indications provided by sampling plans and control charts. Mistakes in rejections still occur, but they are few and far apart. In many instances, the improved uniformity of product, moving from one operation to the next, has permitted a reduction of down-time and increased output.

Examination of outgoing products has revealed striking gains in the appearance of our merchandise: clearer patterns, truer contours and a generally higher quality of finish. These are improvements which cannot be presented in the form of figures, but they have been recognized by the Sales Department. In the long run, this improved quality will make our merchandise more competitive by appealing to the individual shopper and by increasing the acceptance of our product among our dealers, who readily detect changes in quality.

Within the next few months, it is contemplated to introduce additional refinements of methods. In particular, it is planned to set up quality control charts on our drilling and shaping equipment and to install additional sampling plans where needed.

Continuing progress is made towards close cooperation with the Purchasing Department in the elimination of unsatisfactory sources of supply of raw materials and to present essential inspection data to the Engineering Department, which may be useful in initiating changes in machinery or product design or serve as a basis for setting and reviewing tolerance and specifications.

H. Brown, Quality Control Mgr.
K. Clark, Quality Control Engr.

Fig. 11-2.   Report of results accomplished with quality control program.

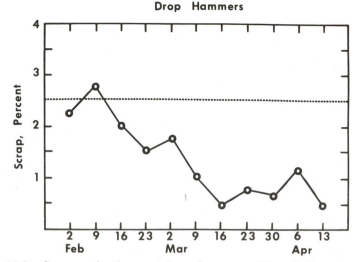

Fig. 11-3. Scrap reduction on drop hammers. Dotted line represents prior average.

It should be realized that in practice there are many times when one of the advantages or disadvantages outweighs all the rest. For example, during an inspection of chrome-plated items, it would be highly impractical to attempt to measure the brightness of finish along a numerical scale. Therefore it is preferable to classify each article "satisfactory" or "defective" and use a sampling plan. On the other hand, when determining yield point and elongation of steel castings, testing is so difficult, costly, and time-consuming that it is of primary importance to have small sample sizes. Here the control chart method must be used.

The Human Factor in Introducing Quality Control

Installation of quality control is not only a technical problem, it is also a problem of human relations. As is only too well known to any quality control engineer, one of the most frequent causes of failure is the resistance built up by the shop man against the outside engineer who has come to establish the new method. Resistance to change is, after all, a normal human reaction. It may be aggravated by fear of loss of standing or even of losing one's job, by suspicion, or by simple stubbornness. Much time may have to be invested in tactful diplomacy to overcome this resistance.

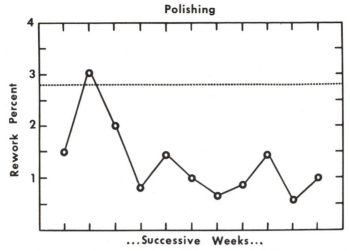

Fig. 11-4.   Decrease in rework in polishing. Dotted line represents prior
average.

In dealing with this phase of quality control, the inside inspection
supervisor is in a far better position than the outside engineer. The
former not only knows his men, machines, and product, but has
already had an opportunity to gain the confidence of both foremen
and operators. This will prove helpful when he begins with his new
project; moreover he is not pressed constantly for time in completing his task, but the outside man, as a rule, is expected to show
results quickly.

Installation Proposal

When the installation of quality control is being considered for a
manufacturing organization, it is often useful to present a proposal
to top management, as a means of soliciting approval and support
and obtaining the necessary financial funding. The actual form of
the installation proposal will vary greatly, depending upon the
industry and the nature of the organization. Nevertheless, it may be
of some value to present a case history, taken from the metalworking field. The example chosen has been taken from the manufacture
of cutlery and flatware, since it represents relatively simple production operations and thus permits us to concentrate on ideas without
getting involved in technological complexities. The report, with
some modifications, is reproduced in Fig. 11-1.

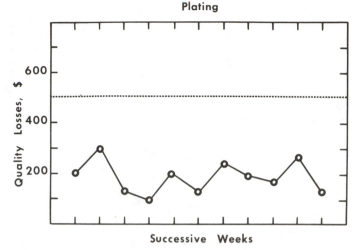

Fig. 11-5. Shrinkage of quality losses on plating process. Dotted line shows prior average.

Installation Review

The initial memorandum recommending installation of quality control should be followed up by a report documenting the achievements attained from the operation of the program. Figure 11-2 presents such a report. Wherever possible, this document should be supplemented by visual presentations. For example, the report mentions savings of scrap and avoidance of quality losses on drop hammers, in polishing, and in plating. Figures 11-3 through 11-5 show these accomplishments in graphic form. Further, Fig. 11-6 gives an illustration of the inspection standardizations effected. It constitutes further evidence of work accomplished that might be included in the report.

REVIEW QUESTIONS

1. Why is the pilot run so important when starting a quality control program?
2. In what places should one look for promising starting points to begin installing a quality control program?
3. It has been emphasized that quality control must be installed with the needs of different areas of application in mind. Can you list some of these areas?
4. What are the advantages and disadvantages of control charts versus sampling plans? (Answer in the form of tabular comparison permissible.)

QUALITY CHARACTERISTICS LIST		
Product: Knife blades	Operation: Mirror finish	Date:
EXAMINE	**MAJOR DEFECT**	**MINOR DEFECT**
Entire blade	Scorch marks	Satin finish
	Glaze marks	
	Scratches	
	Dark burns (compare with standard sample)	Light burns (compare with standard sample)
Back	Burrs	Thin
	Nicks	
Tip	Nicks	Thin
Edge	Uneven taper	
	Thick or thin (check with go—no go gage)	
	Wet grind marks	

Fig. 11-6.   Streamlining and standardizing inspection procedures by means of standard inspection lists.

5. What are the principal human factors to be considered in planning quality control work?
6. Quality control is done not just to improve quality, but also to enhance product marketability and reduce costs. How can these points best be illustrated? (Give a brief example of how one can improve quality and marketability while simultaneously reducing costs as a result of the quality control effort.)
7. What is the purpose of a standard inspection list?

CHAPTER 12

Analysis of Variance

Once we have determined a particular variation, we may find it important to analyze it and break it down into its component individual factors. In production, an observed variation usually consists of a set of smaller variations, much in the manner of a famous De Morgan verse, often quoted by statisticians:

"Great fleas have little fleas upon their backs to bite 'em,
And little fleas have lesser fleas, and so ad infinitum."

Let us examine how this principle works in industrial statistics.

Breakdown of Variations

Reminders that an observed variation often consists of smaller component elements surround us much of the time. For example, when calculating the standard deviation for a production process, we tacitly recognize that the value obtained includes the (hopefully small) variations contributed by imperfections of the measurements on each unit of product. Just as product tolerances can never reach zero, so can the precision of human instrumentation never be perfect.

By use of what is called *variance analysis* we can isolate and identify the magnitude of the "flea" representing the precision of measurement, so that we are then left with the net value of process variability. Going one step further, however, we can pull another "flea" off the variations because of the precision of measurement. In particular, we can isolate the variations due to the human element, associated with the skill and care of the average tester. As the remainder, we then are left with the inherent precision of the measuring equipment itself, be it a dial gage, a scale, or other instrument.

Turning our attention to the process variability itself, we can again break this down into various elements derived from the pro-

171

cessing. Some typical applications will be discussed in this chapter. Based on this knowledge, it is often possible to discover hitherto unsuspected causes of product variability and to initiate improvements accordingly. Before discussing how this can be accomplished in actual practice, some basic concepts of analysis of variance should be understood.

The Concept of Variance

In statistical quality control language, "variance" refers simply to the squared standard deviation. To take an example from a filtration process, if the standard deviation of filtering time for a particular solution is 2.5 seconds per ounce, then the variance is found to be 2.5 times 2.5 or 6.25 seconds per ounce.

The reason for taking squares is that certain calculations are thereby simplified. To explain this, we may for a moment assume that the reader is already familiar with the technique of variance analysis and has found the following standard deviations on a certain machine operation:

Standard deviation inherent in machine 6.4
Standard deviation due to tool wear and setting 4.8
Standard deviation of bulk produced, or overall variation 8.0

All data are in terms of 0.0001 inch. We shall refer to the three types of variation more briefly as *machine, setting,* and *overall deviations,* respectively. Now, if we express these same data in terms of variances, we have:

Type	Calculations	Result = Variance
Machine	6.4 × 6.4	41.0
Setting	4.8 × 4.8	23.0
Overall	8.0 × 8.0	64.0

Note that the total of machine variance plus setting variance, that is, 41.0 plus 23.0, equals the overall variance of 64.0. This fact usually indicates that our data are correct. Had we performed the same addition on the standard deviation figures, we would have had machine deviation plus setting deviation, or 6.4 plus 4.8, which does *not* equal the true overall deviation of 8.0. This phenomenon, which requires the use of squares for addition, is referred to as "Pythagorean" addition, after the ancient mathematician who first discovered this principle when working with triangles.

It is thus shown that if we know two (or more) variances separately, we can very simply find the total or overall variance by adding the separate variances. Accordingly, the term "analysis of variance" rather than "analysis of standard deviations" is used. It should be emphasized, however, that in many practical process investigations we may find it simpler to deal in straight standard deviations or coefficients of variations, so long as there is no need to add them.

Typical Application

Analysis of variance, especially in industrial research and experimentation, can become a very complex subject. Even as applied to ordinary process investigations, it is difficult to find simple examples. A case history, which is believed to demonstrate the basic technique more clearly than any other encountered in the author's experience, is given here. It is taken from a synthetic staple fiber carding and spinning mill, with the following background.

The mill maintained control charts on all major processing operations: carding, drawing, roving, and spinning. These charts showed that, beginning with the drawing process, variability of the product as measured in grain weight per yard was very high. Moreover, this high drawing variation was inconsistent with fairly low variation in the carding department. Why should variation jump so suddenly? An investigation was made into the drawing by checking gear trains, roll settings, weightings, and other adjustments, but no cause could be found. An analysis of variance, applied to the drawing process, was similarly unsuccessful in finding any possible causes.

Now there exists in a spinning mill a condition commonly encountered in many types of semicontinuous processing. The product from one department is mixed in the processing operation of the succeeding department. In particular, the strands of fibers from about 50 carding department machines are all routed through one drawing machine. The significance of this is that an excessive *overall* variation in one department will show up as high *machine* variation in the next department. In our particular investigation we want to know: "Is the excessive drawing machine variation caused by an excessive overall variation in carding?"

To test this, the data in Fig. 12-1 headed "Carding Coefficient

Werner ies – P.O.M. Co.,N.Y.C.	...Carding....COEFFICIENT of VARIATION						Form LC-13a

Mill: —	Dept.: Carding	Mach.: Carding			Prep. by: —	Date: —
Item Tested: Grain Weight per Yard		Stock: No. 3				
Yds. Tested: 4 × 5	Doublings: —	Std.: 55.0	High: 52.0	Low: 58.0	Average: 55.2 grs.	

Date:	6/1	6/1	6/1	6/1	Range (over-all)					Range (over-all)
Mach. No.	43	27	16	5						
	55.5	51.8	52.0	56.0	4.2					
	60.2	54.2	56.1	57.1	6.9					
	54.8	50.1	54.1	55.2	5.1					
	52.1	54.1	53.0	51.1	3.0					
Range (within)	5.3	4.1	4.1	6.0						

Date:	6/2	6/2	6/2	6/2	Range (over-all)					Range (over-all)
Mach. No.	8	39	7	12						
	50.3	59.9	55.5	58.8	9.6					
	58.8	58.2	50.2	56.1	8.6					
	54.0	56.3	51.6	59.2	7.6					
	53.8	57.0	54.7	57.1	3.3					
Range (within)	6.5	3.6	5.3	3.1						

Date:	6/3	6/3	6/3	6/3	Range (over-all)					Range (over-all)
Mach. No.	21	29	41	32						
	50.4	55.4	58.8	58.6	8.4					
	52.1	56.3	57.3	59.2	7.1					
	51.6	56.1	56.8	55.7	5.2					
	53.5	56.1	53.9	57.1	4.6					
Range (within)	3.1	2.7	4.9	3.5						

SUMMARY	Within	Overall	Notes:
1. Total of Ranges	54.2	73.6	
2. Average Range	4.5	6.1	
3. Average Range, %	8.2	11.1	
4. Coefficient of Variation	4.0 %	5.4 %	
5.			

Fig. 12-1.   Carding coefficient of variation.

of Variation" were obtained. From twelve randomly selected machines, samples of four strands each were taken, weighed, and written down in terms of grains-per-yard. The range-within was then found for each machine, such as 5.3 for machine No. 43, 4.1 for machine No. 27, and so on. The total of all these ranges is 54.2

grains; divided by twelve, it gives an average range-within of 4.5 grains. Expressed as a percentage of average grain weight of 55.2, this range becomes 8.2 percent. Then, by using our well-known conversion factor for samples of size four, 0.49, we find the coefficient of variation to be 4.0 percent.[1] These are familiar computations, and the resultant variation is the machine variation, also known as "within-machine" or "within-sample" variation, and the type which is usually represented by range control charts.

To find the overall variation, we should now take our 48 test results (twelve samples, four tests per sample), write each on a little slip of paper, and throw them all into a bowl or hat. Then we should draw individual slips, with each group of four representing a new "sample." Taking ranges of these groups of four, we would then be able to determine the overall variation of the department. Now in our particular example, both the machines and the individual lengths per sample tested were taken at random and without any significance as to sequence. Therefore, the data are already well enough mixed to obviate the necessity for hat-drawings. Instead, we merely take the range-across, giving us the overall variation. For example, the first line of test results lists, for machines numbered 43, 27, 16 and 5, the data: 55.5 grains, 51.8 grains, 52.0 grains, and 56.0 grains. The range of these is 56.0 minus 51.8 or 4.2, as shown on the illustrative form. Proceeding in a similar manner for the next line, we find a range of 6.9 grains. Then, finally, we find a total of ranges equal to 73.6 grains, yielding an average range of 6.1 grains or 11.1 percent (in terms of the average grain weight of 55.2) and a coefficient of variation of 5.4 percent. Summarizing, we have found:

Coefficient of Variation, Within Machines 4.0
Coefficient of Variation, Overall 5.4

Clearly, the overall variation is considerably above the within-machine variation and must be investigated as the cause of the high variation shown on the control charts for drawing. The actual fault was found to be use of nonstandard tension gears, which produced

---

[1]Note that the factor F_r from Table 6-2 permits two types of conversion. The first is from average range $\overline{R}$ to standard deviation σ. The second is from average range, *in percent*, $\overline{R}\%$ directly to *coefficient of variation*. In particular, $\overline{R}\%$ is the average range expressed as a percent of the grand average. Similarly, the variation coefficient is the standard deviation expressed as a percent of the grand average. Multiplying $\overline{R}$ by F_r yields the estimated standard deviation, while multiplying $\overline{R}\%$ by F_r yields the *coefficient of variation*.

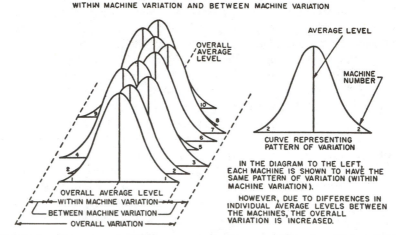

PROCESS VARIATION IS COMPOSED OF TWO MAJOR PARTS –
WITHIN MACHINE VARIATION AND BETWEEN MACHINE VARIATION

Fig. 12-2.   Process variation is composed of two major parts, within-machine
variation and between-machine variation.

excessive between-machine variation, not computed here. The total
of the within-machine and between-machine variations constituted
the high overall variation.

Within-Machine and Between-Machine Variation

Wherever there is a processing department in which several
machines produce the same material, we must make sure that both
within-machine and between-machine variations are kept under
control. If we fail in this, then the overall departmental variation
will be excessive. How this works is illustrated in Fig. 12-2.

On the right-hand side of this illustration, we show a typical
frequency distribution curve representing the variation pattern of a
machine. The center line in the middle represents the average level.
Now let us assume that we have several machines in a department,
and that this frequency curve represents the average within-machine variation. For the preceding example, this would mean a coefficient of 4.0 percent. However, due to the fact that the same tension gears are not used on each machine, there is variation in the
amount of drafting between the different machines. This results in
differences between machine levels, or between-machine variation.
The left-hand side of our illustration shows how, by displacement of
the average level, the individual machines deviate to the right and
left. For example, machines No. 3 and No. 7 are far to the right of

center, and machines No. 4 and No. 9 are far to the left of center or overall average level. This produces the additional variation shown as "between-machine variation" on the chart. The overall variation is then the combined total of the within- and between-machine variations.

Where control charts are maintained on individual machines, this situation of excessive overall variation would not show up in the particular department involved. It does reveal itself, however, in that process where the product from several machines in a preceding department is mixed, blended, or otherwise combined.

Our last illustration clearly shows that it is a physical impossibility for overall variation in a department to be ever truly less than within-machine variation. Yet, sometimes you may get a figure for overall variation that is below that for within-machine variation. When this occurs, several possibilities are present:

1. An error may have been made in calculations.
2. The number of samples is inadequate. As a rule you should use enough samples to fill out the entire form in Fig. 12-1. (For simplicity of illustration, only 12 of the possible 24 samples were used in the example, which is still permissible but not the safest practice.)
3. Where samples represent routine mill testing results, and possibilities (1) and (2) have been ruled out, we must suspect the testing performance. Under pressure of time or for other reasons a tester who skips some tests might usually be good at guessing correct averages, but he would have to be a genius to make the variations jibe as well.

Evaluation of Variations

When overall variation exceeds within-machine variation, this does not necessarily mean that the difference observed is a true one. After all, we have computed the variation coefficients from sampling data, and the observed difference in the variations may be caused by chance fluctuations of sampling and testing. We thus need a method of evaluating the difference in the variations observed, and to judge whether the excess of overall over within-machine may be considered real or ascribable to possible chance fluctuations of sampling and testing. Such an evaluation procedure is known as a "significance test." This test may be described by the simple steps shown below, using for illustration the data obtained from Fig. 12-1:

1. Form a ratio of overall to within-machine variation. Thus, for the values of 5.4 percent and 4.0 percent for overall and within-machine variation respectively, the ratio is 5.4 percent $\div$ 4 percent or 1.35.
2. Observe that the *sample size* or, in other words, the number of test values used for *each* within-machine range is 4. The *number of samples*, representing the number of groups from which ranges were obtained, is 12. (This is the same 12 which we used to find the average range.)
3. Enter Table 12-1 for sample sizes of 4 and the number of samples, which is 12. This shows that the minimum ratio needed for significance is 1.133. Since the actual ratio, from step 1, is 1.350 and thus greater than the minimum needed, we may consider the excess of overall over within-machine variation to be *significant* (and thus *not* ascribable to chance fluctuations).

In many cases, as in Fig. 12-1, the differences between the variations may be so large that no formal significance test is required. Inspection of the results alone will suffice. However, it is often dangerous to rely on such off-hand judgment. Also, where differences are not so clear-cut as in the example, only the ratios in Table 12-1 can serve as a useful criterion.

As in all sampling, the significance test furnished here involves a certain sampling risk. In particular, the values in Table 12-1 have been so computed as to involve a risk of 5 percent of calling an observed difference "significant" when actually it is caused by chance fluctuations of sampling and testing. This appears to be a practical risk for this type of work.

Between-Machine Variation

The methods shown above involve the direct determination and comparison of variation coefficients representing within-machine and overall variation. The between-machine variation can be found indirectly from these two values, by means of the "Pythagorean" addition of variances, shown under "Concept of Variance" in this chapter. Based on these calculations, the convenient nomograph in Fig. 12-3 has been prepared, from which the between-machine variation coefficient can be read on the basis of the particular within-machine and overall variations observed.

Table 12-1.   Values of the Minimum Ratio of Variation Coefficients Needed for Significance

Sample Size	Number of Samples									
	10	12	14	15	18	20	25	30	40	50
2	1.453	1.393	1.351	1.334	1.293	1.272	1.234	1.208	1.173	1.145
3	1.223	1.197	1.178	1.170	1.151	1.141	1.123	1.110	1.093	1.081
4	1.150	1.133	1.121	1.115	1.103	1.097	1.084	1.076	1.064	1.056
5	1.113	1.101	1.092	1.088	1.078	1.074	1.064	1.058	1.049	1.043
6	1.091	1.081	1.074	1.071	1.063	1.060	1.052	1.047	1.040	1.035
8	1.066	1.059	1.053	1.051	1.046	1.043	1.038	1.034	1.029	1.026
10	1.052	1.046	1.042	1.040	1.036	1.034	1.030	1.027	1.023	1.020

SOURCE: From the author's paper, "Variations Flow Analysis," *Technometrics*, vol. 2, no. 3 (August, 1960): 373–386.

Example: Given 10 samples, each consisting of 4 tests. To find ratio, enter table at the level of sample size = 4. Proceed horizontally until column *number of samples* = 10 is reached. In the cell so found, the ratio 1.150 is obtained.

NOTE: Significance is at 95 percent confidence, and thus entails a 5 percent risk of error.

Basis of Nomograph:

$$(\text{Within-Machine Variation Coefficient})^2 \; + \; (\text{Between-Machine}$$

$$\text{Variation Coefficient})^2 \; = \; (\text{Overall Variation Coefficient})^2$$

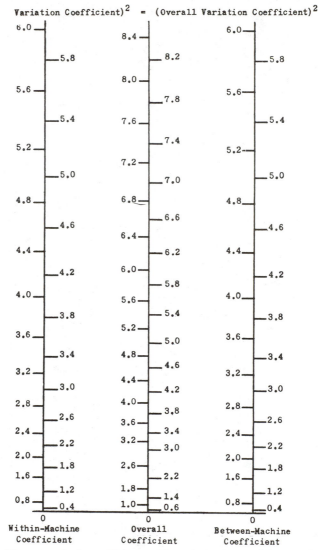

Fig. 12-3. Nomograph for within-machine, between-machine, and overall variation coefficients. *Example*: For a coefficient-within of 3.0 and a coefficient-overall of 5.0, draw a straight line between these two points on the first and middle scales. The extended line cuts the between-coefficient scale at 4.0, which is the answer.

Source: V overall $= \sqrt{V^2 \text{ within} + V^2 \text{ between}}$, where $V =$ variation coefficient.

Using Analysis of Variance

Unlike control charts or sampling plans, analysis of variance is not the type of tool that can be used routinely in a plant. It is a technique for analyzing and disentangling a series of individual variances, and much thought must be given to the setting up of the test arrangements and the interpretation of the data.

It is hard to draw the line as to which analysis of variance uses fall properly within the province of statistical quality control, and which belong in the realm of industrial research and experimentation. In this writer's opinion, the quality control engineer need not be familiar with the complexities, details, and ramifications of the method. As long as he is aware of the basic essentials involved in the analysis of variance approach, as well as the type of information it can furnish, he will usually be sufficiently familiar with this technique. Then, when a special problem does come up where analysis of variance can be seen as desirable, specialized statistical assistance can be obtained both for the design of the testing program and the analysis of the data to yield information that may have meaning with regard to processing improvements.

What has been presented in this chapter is a simplified yet practical technique for elementary application of analysis of variance as a quality control tool in the industrial plant.

REVIEW QUESTIONS

1. Why would one wish to perform an analysis of variance? (Give an example.)
2. Given a standard deviation overall of 5 and a standard deviation within of 4, what would be the standard deviation between groups? (Show the formula, then the computations, and finally check your result with the nomograph provided in the chapter.)
3. In general terms, why would it be *incorrect* to answer the question as follows:

   Standard Deviation Overall 5.0
   Standard Deviation Within Groups 4.0
   Standard Deviation Between Groups
   $= 5.0$ minus 4.0 .... 1.0
4. Is analysis of variation substitutable for control chart methods, and if so, under what conditions?
5. What is Pythagorean addition, and how is this concept useful in analysis of variance?
6. In the assembly of covers to pump meter bodies, the torque readings

in kilograms for the three capscrew positions A, B, and C were obtained:

June 8

A	B	C
4	2	6
3	1	4
5	3	2

June 10

A	B	C
6	4	1
5	4	3
4	4	2

June 11

A	B	C
8	5	5
6	4	5
4	6	7

June 12

A	B	C
3	3	6
4	6	5
3	8	4

What are (A) process average; (B) within-position, between-position, overall standard deviations and (C) within-position, between-position, and overall variation coefficients?

See also Cases 7 and 8, Appendix 3, for additional review material.

CHAPTER 13

Optimizing Processing
Through Evolutionary Operation

Evolutionary operation is a special technique for the step-by-step improvement of a process until it reaches an optimum level, resulting in enhanced quality, lowered costs, and increased yield.[1] The method is applied to daily production by introducing a series of systematic small changes in the levels at which process variables are held. After each set of changes, the results are reviewed and new changes made—the goal of this evolutionary development being to "nudge" the process gradually into optimum operating levels. A simplified example, taken from the sizing of plied synthetic filament yarns, will illustrate how evolutionary operation works.

Example of Evolutionary Operation

The primary purpose of the sizing operation is to condition the filaments for weaving so as to minimize yarn breakage and hence maximize weaving efficiency. Existing mill practice involved settings yielding 6 percent moisture content and 2 percent elongation of the yarn, which is shown as setup 1 in Fig. 13-1. It was decided to check this condition against the additional setups 2 to 5 indicated on the diagram. These further conditions represent small changes, which were considered adequate to produce discernible effects on weaving efficiency, yet small enough to assure that no undue production or quality problems would arise.

Each set of operating conditions was run four times, resulting in four cycles, with weaving efficiencies as shown on the next page.

---

[1] Developed by G. E. P. Box, Imperial Chemical Industries Limited, first published in *Applied Statistics,* vol. 6, no. 6, 1957.

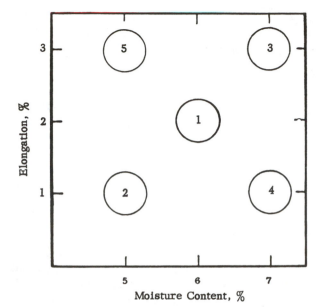

Fig. 13-1. Design of investigation, comparing processing setup 1 against setups 2 to 5. Setups are shown in circles.

	Setups				
	1	2	3	4	5
	Efficiencies, Percent				
Cycle I	91	93	94	98	92
II	94	95	95	97	93
III	94	92	95	96	92
IV	93	96	96	93	91
Average	93	94	95	96	92
Range	3	4	2	5	2

The average of the ranges is 3.2. Since each of the setup averages is based on four determinations, we enter at this level Table 7-1 to obtain under column d the control limit factor of 0.49. Therefore, ± 0.49 times 3.2 or ± 1.6 represents the extent of two standard errors[2] around the setup average, in percent weaving efficiency. We may call this ± 1.6 more simply the "error limit."

---

[2]The reasons for using two standard errors here are the same as those for control chart purposes, as discussed in Chapter 7.

The relative effects of moisture and elongation are estimated from the arrangement below, in which the setup averages are recorded to correspond to the relative positions of the setups given in Fig. 13-1.

Elongation Percent	Moisture Content, Percent 5	6	7	Average	Effect of Elongation
3	92		95	93.5	
2		93			— 1.5
1	94		96	95.0	
Average	93.0		95.5	94.25	

Effect of
Moisture   + 2.5

The effect of elongation, −1.5 in percent weaving efficiency, is found by subtracting the elongation average of 95 from the elongation average 93.5. By similar subtraction of 93.0 from 95.5 for the moisture averages, the effect of +2.5 of moisture content is found.

The observed effect of moisture, +2.5, is well in excess of the error limit, ±1.6, and indicates strongly that increased moisture content of the yarn is beneficial for weaving efficiency. The observed effect of elongation, −1.5, is slightly below the error limit, but still large enough to suggest that reduced elongation is beneficial. In general, an observed effect which is below the error limit but still exceeds half its magnitude may still be considered suggestive of a true effect. However, the smaller the observed effect in relation to the error limit, the greater is the probability that it may represent merely chance fluctuations of sampling and testing.

It is also interesting to compare the efficiency of setup 1, 93 percent, against the average of 94.25 percent for setups 2 to 5, which shows that the evolutionary operation just completed in its first phase has benefited and not harmed efficiency. A further evaluation may be obtained of the effect of interaction of moisture and elongation, by subtracting the average of setups 4 and 5 from the average of setups 2 and 3. In particular, 94.5 minus 94.0 yields 0.5, which is a negligible effect by comparison against the error limit of ±1.6.

We are now ready to decide on the next phase of evolutionary operation. Based on the observed effects, a good decision for further evolutionary changes in processing would be as follows: run a further phase of four cycles; select the center point, representing the new setup 1 at a somewhat higher moisture content of 6.5 and slightly reduced elongation of 1.5 percent; and set up further processing conditions 2 to 5 at spacings of 1 percent elongation and 1 percent moisture in a symmetrical pattern parallel to that of Fig. 13-1, but centered around the new setup 1. Then, after this new phase of operations has been completed, the results can be reviewed again and further phases planned accordingly.

Once experience has been gained with this type of operation, the program may be expanded. Additional criteria of processing cost and product quality may be included. Further processing variables, such as roll pressures, speeds, and drying rates, may be incorporated in the investigations.

Multivariate Programs

Evolutionary operation involving three input variables is best portrayed in a three-dimensional view, as in Fig. 13-2. An electronic soldering operation was run at a central level, while three input variables of pulse time, current power, and clamping force were held at slightly higher (+) and also slightly lower (−) levels in successive trials. Output, in terms of the shear strength of solders in kilograms, is entered in the nine circles (eight corners and one central point). Assuming that the outcomes are significant, it is clear that in another phase a new center should be investigated with higher levels of time and power. Results with regard to clamping force are

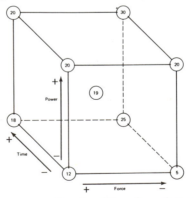

Fig. 13-2. Evolutionary operation involving four factors. Input factors or variables are pulse time, current power, and clamping force in an electronic soldering operation. Output variable is shear strength of solder in kilograms. Levels at slightly above (+) and slightly below (−) center point (with output of 19 kilograms) are studied.

somewhat contradictory—in some instances higher force yields less strength; in others it yields more strength. Further study of this phenomenon will be required. It is also apparent that three input variables represent the maximum that can be introduced in one phase while still attaining a good overview of results. For four input factors it would be impossible to present a comprehensive graphic presentation. It is, of course, feasible to show several cubes for outputs. As an illustration, had we been interested not only in strength but also solder consumption in grams, we would merely have provided a further cube with circles representing the outputs in grams per, say, 100 solder connections.

Laboratory Optimization

It might be argued that evolutionary operation during production is relatively slow, and that optimum levels of variables might best be established in the research laboratory. Unfortunately, however, it is usually impossible in the laboratory to duplicate plant conditions satisfactorily and the optimum level of actual production will therefore differ from the theoretical level predicted. The laboratory works out new processes and procedures and establishes levels which serve as starting points. Evolutionary operation then investigates successive modifications of the variables, with the objective of bringing the process to optimum levels.

Practical Philosophy

Since evolutionary operation, or EVOP, is relatively simple to apply, it has been made part of daily production in quite a number of instances. The manufacturing process thus does not only yield product, but also information on cost, productivity, and quality improvements. A revealing testimonial on the practical values of this approach was given by R. S. Riordan, Jr., Superintendent of Process Improvement, The Chemstrand Corporation.[3] Excerpts appear below.

"After roughly 15 to 16 months experience, it is possible to make a fair appraisal of Evolutionary Operation. Contributions will be classified under three headings: financial, technical and personnel.

---

[3]Presented at Gordon Research Conference on Statistics, August 24, 1959 and quoted in N. L. Enrick, *Sales and Production Management Manual* (New York: Wiley, 1964).

"Financial. Using direct savings in raw materials, we have had programs which ran for long periods and saved only $10,000 per year; we have also been fortunate in having programs which saved $100,000 per year. Not all savings have been in raw materials—others were in increased capacity, lessened downtime, and the like.

"Technical. We have never had an Evop program that did not teach us something. This flow of technical knowledge has given us information leading to better understanding of the operations.

"Personnel. Evop has been very effective in human terms. Two years ago, for instance, we had one area where work and morale were at a minimum. It just seemed impossible to do anything that had not been done already. Today, with Evop, you have to stand in line to get a production unit on which to run your program. There is not enough equipment for all the tests we have in mind. The men are working much more effectively and they are many times as happy. Morale could not be better."

While gains in terms of increased technological understanding and higher morale cannot be precisely measured in dollars, they are nevertheless an important benefit of a successful program.

The type of application will vary. While we have presented an illustration involving two input variables (moisture and elongation) and one output variable (yield), many plants prefer to use three input variables and to observe several outputs (such as quality, yield, and productivity). In all instances it may be best to begin with a simple program, using a minimum of variables, and gradually progress to greater depths as the people concerned learn to work with the new techniques.

REVIEW QUESTIONS

1. What benefits are attainable with evolutionary operation?
2. You wish to determine the optimal combination of two additives, A and B, in the production of an extruded plastic. The levels under consideration are 0.1 and 0.3 percent for A and 0.4 and 0.8 for B. Draw a diagram of what the investigational setups should be.
3. For the problem just discussed, at what levels of A and B will setup No. 1 occur?
4. What is the difference between setups, phases, and cycles in evolutionary operation?
5. Are we necessarily confined to the study of just two factors?
6. Why is laboratory optimization often not as efficient as evolutionary operation?

See also Case 12, Appendix 3, for additional review material.

CHAPTER 14

Statistical Build-up of Tolerances

Suppose that we are manufacturing three gears of 1-, 2-, and 3-inch face widths, respectively, and that the gears are to be assembled next to each other on a transmission case shaft. If the overall length of the assembled gears must be held within a tolerance of ±0.018 inch, must the tolerance on the width of each of the three gears be held to ±0.006 inch or can a larger tolerance be permitted?

The answer to this depends, among other things, upon the background of the person questioned. A novice in manufacturing practice might say, "To be absolutely sure the assembly will not exceed the ±0.018 inch tolerance, it is necessary to add the tolerances on the individual widths; therefore, ±0.006 is the largest tolerance we can afford to give the individual parts since $3 \times \pm0.006 = \pm0.018$."

On the other hand, a more experienced observer might say, "It seldom happens that three different parts are all at their maximum (or minimum) dimension at the same time, and it is therefore probable that a tolerance greater than ±0.06 can be applied to each part without exceeding the assembly tolerance of ±0.018 inch."

It may be seen from these replies that the novice is disregarding extra manufacturing cost and playing 100 percent safe by specifying a tolerance of no more than ±0.006 inch while his more experienced friend is willing to gamble that most of the time a tolerance greater than ±0.006 inch will not result in more than ±0.018 inch in the assembly.

However, this latter viewpoint still leaves unanswered the query, "How much more than ±0.006 inch can the tolerance be?" It is the purpose of this chapter to provide some of the answers to this kind of question.

Establishing and meeting tolerances are problems closely interwoven with all inspection, testing, and control work. We have previously demonstrated the need to evaluate process capability so as to assure that the production operations can routinely and economically satisfy a set of specifications. Control charts for sample averages—using limits consistent with process variability, sampling variations, and the tolerance range—were discussed at length. In the following, our analysis is carried one step further. We shall consider the manner in which tolerances build up when components must mesh in chainlike or other interlocking and interdependent configurations.

Application of Variance Analysis

While the procedures described in Chapter 8 were concerned primarily with the evaluation of *individual* tolerances, the present chapter shows how tolerances may be *combined*—other than by simple addition and subtraction—to satisfy a required overall tolerance in assembly. In this work the principles of addition of variances discussed in detail in Chapter 12 may be applied with the result that the final tolerance values established by means of these techniques will permit maximum product variation consistent with quality and processing requirements. Thus, statistical methods can help us to obtain the cost-saving advantages of maximum allowable tolerances consistent with the requisite quality.

Variances in Accumulative Assembly

When a dimension in an assembly represents a linear combination of several components, it is apparent that the variance in the total dimension is the result of the variances of the parts. Consequently, the relationship between total and individual part variances may be evaluated by means of the statistical techniques of analysis of variance.

Example 14-1.   Linear Combination of Components and Resultant Standard Deviation Obtained from Gear Assembly

	Gear A	Gear B	Gear C	Assembly
	(Thousandth of an Inch)			
a. Basic Size	1,000	2,000	3,000	6,000
b. Standard Deviation	2	2	2	*
c. Variance ($= b^2$)	4	4	4	12
d. Standard Deviation of Assembly $= \sqrt{12}$				3.46

*Note that standard deviations cannot be added directly. Instead the analysis of variance, steps in lines c and d are needed to find the assembly standard deviation.

To illustrate, if the overall dimension of three adjoining gears on a transmission shaft is comprised of three widths of 1, 2, and 3 inches, and each component has been shown from control charts to be normally produced with a standard deviation of 0.002 inch in the width, then the expected standard deviation of the combined dimension, under random assembly, may be determined as shown in Example 14-1.

The steps shown in Example 14-1 may be expressed in a formula:

$$\text{Assembly Standard Deviation} = \sqrt{\text{Sum of Squares of Individual Standard Deviations}}$$
$$= \sqrt{2^2 + 2^2 + 2^2}$$
$$= \sqrt{12} = 3.46 \tag{14-1}$$

The standard deviation of the combined linear dimensions of the widths will become important when considering the interaction with, for example, a bearing housing on one side and a shaft shoulder on the other.

The data are in terms of 0.001 inch. Formula 14-1, familiar from analysis of variance, is also fundamental to statistical tolerancing. When the standard deviations of the individual parts are identical or approximately alike, the equation simplifies to:

$$\text{Assem. Std. Deviation} = \text{Indiv. Std. Deviation} \times \sqrt{\text{No. of Parts}} \tag{14-2}$$

Thus, for Example 14-1, using units of 0.001 inch,

$$\text{Assembly Standard Deviation} = 2 \times \sqrt{3} = 3.46$$

This simplified formula does not apply where the standard deviations of the components differ. Thus, where the individual standard deviations are 3 and 4, respectively, we would use Formula 14-1 to get:

$$\text{Assembly Standard Deviation} = \sqrt{3^2 + 4^2} = 5$$

The formula is also valid when variation coefficients are used, since a variation coefficient is merely the standard deviation expressed as a percentage of the distribution average. Moreover, the answer just found could have been obtained directly from the nomograph in Fig. 12-3, by drawing a line between the values 3 and 4 of the within-machine and between-machine coefficients columns, respectively, and reading 5 from the overall coefficient column.

Requirements for Formula to Work

The relationship of variances in random assembly, shown by Formula 14-1, will be strictly valid only when the individual parts come from processes under good control, as established from a review of control charts. Such a production process is said to be "in statistical control." It is also desirable that frequency distributions derived from the control chart data be checked, to see that they fall into the approximate pattern of a normal distribution.

Where parts do not come from a process in statistical control, but are thoroughly mixed prior to assembly to assure good randomness, Formula 14-1 will still be applicable. It has also been demonstrated that the validity of the formula will not be seriously affected when distribution patterns are either much flatter or else much more peaked than the bell shape of the normal curve, shown in Fig. 6-3, and approach instead an almost rectangular or triangular form. Thus, the relationship depicted in this section should have wide application. Special cases in exception to the general rule will be discussed later.

Tolerances in Accumulative Assembly

In the same way that we can add standard deviations to obtain, by means of the analysis of variance formula, an estimate of the overall standard deviation, we can also add individual tolerance values to find overall tolerances. This becomes apparent when one considers that the tolerance is merely another way of stating the allowable standard deviation of product. In particular, when a tolerance is set at ± 6, then dividing this quantity by $\pm$ the factor 3 (from Table 6-1 of page 75) will yield 2 as the highest allowable standard deviation that can be tolerated without producing defectives.[1] Conversely, if the standard deviation is 2 for a product obtained from a process under statistical control, then the tolerance which can be maintained by this process is $\pm 3 \times 2$ or ± 6. This tolerance represents the capability of the process and is generally called the "natural tolerance."

Using the data from Example 14-1, we may now show how the

---

[1]Sometimes a more precise factor of 3.1 is used, as discussed in Chapter 6. Usually, however, the more convenient rounded value of 3 is considered adequate, since errors in estimating the standard deviation and departures from normality are generally expected to be greater than this small difference of 0.1 anyway—without affecting the practical validity of the factor.

same analysis of variance formula may be adapted to apply also to tolerances. For this purpose Example 14-2 is provided. In the example it is assumed that the tolerance specified by the designer is approximately equal to the natural tolerance of the process. Note that steps d and e represent the same analysis of variance procedure as in Example 14-1, but this time applied to tolerances. Moreover, if the assembly standard deviation of 3.46 from Example 14-1 is multiplied by the tolerance factor of ± 3 from Table 6-1, page 75, we obtain the same assembly tolerance of ± 10.4 shown in Example 14-2. This illustrates the consistency of the two methods.

The formula for assembly tolerances is thus:

$$\text{Assembly Tolerance} = \sqrt{\text{Sum of Squares of Individual Tolerances}} \qquad (14\text{-}3)$$

In Example 14-2,

$$\text{Assembly Tolerance} = \sqrt{6^2 + 6^2 + 6^2} = \pm 10.4$$

Again, since the tolerances of the individual parts are all alike, we may simplify:

$$\text{Assembly Tolerance} = \pm 6 \times \sqrt{3} = \pm 10.4$$

Had we used four parts in place of three, all alike as to individual tolerances of 6, we would have had:

$$\text{Assembly Tolerance} = \pm 6 \times \sqrt{4} = \pm 12$$

Example 14-2.   Linear Combination of Components and Resultant Tolerance, Obtained from Gear Assembly

	Gear A	Gear B	Gear C	Assembly
	(Thousandths of an Inch)			
a. Basic Size	1,000	2,000	3,000	6,000
b. Standard Deviation	2	2	2	3.46
c. Tolerance $(= \pm 3 \times b)$	± 6	± 6	± 6	*
d. Squared Tolerance	36	36	36	108
e. Assembly Tolerance $(= \sqrt{108})$				± 10.4

*Note that tolerances, like standard deviations, cannot be added directly. Instead, the analysis of variance steps in lines 4 and 5 are needed. The standard deviation of 3.46 in line b was obtained from Example 14-1.

Example 14-2 assumed the specified tolerance to be approximately equal to the natural tolerance; the formulas shown would work just as well in cases where the specified tolerances are larger than the natural tolerance, so long as each individual tolerance is enlarged by approximately the same proportion. When specifications are wider than existing process capabilities, less frequent machine adjustments, omission of a processing stage, or use of a learner in place of a skilled operator may be considered.

Tolerances for Individual Parts

In the preceding sections we have considered the overall tolerance obtained for an assembly, based on the tolerances of the individual parts. Often, however, the problem is reversed: The overall tolerance for an assembly is known, and it is desired to determine compatible tolerances for the individual parts. In such cases the individual part tolerance can be found from Formula 14-4.

$$\text{Part Tolerance} = \text{Assembly Tolerance} \div \sqrt{\text{No. of Parts}} \quad (14\text{-}4)$$

As an illustration, for a specified overall dimension of ± 5 thousandths of an inch involving four individual component parts, we have:

$$\text{Part Tolerance} = 5 \div \sqrt{4} = 2.5 \text{ Thousandths}$$

This formula assumes that all individual part tolerances can readily be made alike. Often, however, this is not the case, as will be illustrated in the example of a plastic coating operation in the next section, showing how an individual tolerance is determined when the overall tolerance and one of two parts tolerances are given.

Example 14-3. Component Tolerance for a Relay

The so-called "pile-up" of insulators and springs in the relay of Fig. 14-1 provides a further illustration of the proper determination of component tolerances. From design considerations, an overall tolerance of ± 0.004 inch had been established for the six insulators. Naively, one might now specify that each insulator should conform to $\pm 0.004 \div 6$ or ± 0.00067 inch. More realistically, a statistical evaluation leads to a wider, equally feasible but more economical parts tolerance of $\pm 0.004 \div \sqrt{6} = 0.004 \div 2.45 = \pm 0.0016$ inch.

For the five springs, the overall tolerance was ± 0.0012 inch. By means of the statistical build-up formula, we determine that for each spring the required tolerance is $\pm 0.0012 \div \sqrt{5} = \pm 0.0012 \div 2.24 = 0.00054$ inch. A naive analysis would have yielded $\pm 0.0012 \div 5 = \pm 0.00024$ inch, at considerable increase in production or purchasing cost for the component.

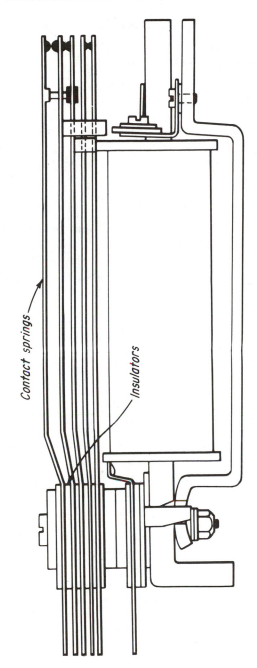

Fig. 14-1.   Relay assembly. A "pile-up" of 5 springs and 6 insulators is portrayed.

Nondimensional Tolerance Accumulation

We may now examine how the formula for tolerance accumulation works in nondimensional applications:

Electrical Resistance. In the assembly in series of four resistances, each having a tolerance of 3 Ohms, the resultant combined tolerance is $3 \times \sqrt{4}$ or ± 6.

Plastic Coating Operation. The tolerance for a coated fabric was ± 5 tenths of an ounce per yard. The finishing plant was able to maintain a tolerance of ± 3 for the weight of coating. Therefore, the tolerance for the uncoated material is:

$$\text{Material Tolerance} = \sqrt{5^2 - 3^2} = \pm 4 \text{ tenths oz./yd.}$$

The value ± 4 represents the maximum permissible tolerance for the uncoated material, which must be specified by the finishing plant, if off-standard product is to be avoided.

Airborne Component. Tolerances for weight of an airborne component, consisting of 14 parts, are as follows:

1. A tolerance of ± 6 oz. for 3 parts
2. A tolerance of ± 8 oz. for 5 parts
3. A tolerance of ± 4 oz. for 6 parts

The assembly tolerance is now:

$$\text{Assembly Tolerance} = \sqrt{(3 \times 6^2) + (5 \times 8^2) + (6 \times 4^2)}$$
$$= \pm 22.9 \text{ oz.}$$

The relationships demonstrated in the various examples in this chapter will work, provided the statistically controlled conditions previously outlined are fulfilled within the practical limitations discussed.

The Old-Timer's Objections

The approach to tolerancing shown in this chapter, using the computation steps borrowed from analysis of variance technique, is relatively recent in industry. It has been a natural outgrowth of the use of statistical control charts and related analysis tools. As a matter of fact, the use of the statistical tolerancing approach shown here presumes statistically controlled processing. The rewards for this care and statistical study come in the form of individual tolerances that can be permitted to be wider than otherwise possible, without detracting from the quality of the finished product. This saves production costs. Moreover, when these techniques are used in analyzing fits and clearances, it can often be shown how tolerances can be widened and quality improved at the same time.

The old-timer, however, who has used the conventional additive methods of tolerance accumulation, may well question our analysis of variance procedures, arguing like this:

"In the example of tolerance accumulation of three individual parts, each being allowed limits of ± 6 thousandths, you used a formula whereby the final assembly would be expected to be within ± 10.4. Yet, if three parts of $+6$ or three parts of -6 are combined, the limits will become ± 18. This is a lot wider than the ± 10.4 from your formula. Therefore, if parts are at their extremes, the assembly limits would be wider than you predict."

It is true, of course, that when the extremes happen to come together, they will assemble to limits in excess of those given by the square root formula. They will then form the simple additive relation shown in Chapter 8. However, under random assembly such a meeting of extremes may be expected only rarely. After all, extremes form only a very small portion of a bell-shaped frequency curve, and one cannot expect that these few parts will normally meet in assembly. A few such adverse combinations will of course occur. Where distributions are normal and assembly is at random, there will be about three out of every thousand assemblies where the statistically determined tolerance is exceeded in assembly. In practice, because of various inaccuracies and imperfections in requsite conditions, the actual occurrence of assembly failures may be several times greater than the theoretical 3 out of 1,000 or 0.3 percent. Usually, however, this presents no real problems in production, since the operator can break up the adverse combination by simply exchanging the extreme piece picked up last for another random piece from the bench.

While the idea of statistical tolerancing was considered quite revolutionary only a few years ago, there are many plants today that find these techniques essential for their work. It is, of course, the responsibility of the statistical tolerancing committees in manufacturing plants to recognize those conditions where the prerequisites for the variance analysis formula are not fulfilled and where the conventional additive or other procedures must be used instead. This will be discussed at the conclusion of the next chapter.

Installation Procedure

The first few applications of any program are usually critical. A new approach is being undertaken, and its validity is in the minds of many, "on trial." Should there be an early failure, it will be hard to overcome the lost confidence. It is thus important to make sure that any statistical tolerancing application introduced, especially at the beginning of a program, occurs under favorable conditions. The following steps should be taken:

1. Ascertain that the production processes are under "statistical control." For this purpose, a control chart should have been run for at least 20 samples. Both averages and ranges should be plotted. If no more than one point is out of control, we may assume satisfactory conditions.
2. Check that each component part represents relatively normal distributions. When from 50 to 100 data points are available, a frequency plot will give adequate indications. For fewer values, normality may be tested with the aid of normal probability paper (such as discussed in Chapter 17 in connection with Table 17-2 and Fig. 17-1).
3. Finally, plot the distribution of the assembled units. It should not only be normal, but also fall within tolerances.

In an organization where quality control is a routine activity, the basic data needed for this analysis are, of course, on hand or readily available.

REVIEW QUESTIONS

1. What analysis of variance principles are used to understand tolerance build-up in assembly operations?
2. For linear combinations, show the formula for the assembly standard deviation.
3. What requirements must be fulfilled if the formula just given is to work?
4. Is there any difference between dimensional and nondimensional linear tolerances? (Give reasons for your answer.)
5. What is the so-called "old-timer's problem" in accepting statistical tolerancing? What factor(s) does the old-timer overlook?
6. Performance characteristics, in percent, for a film resistor have been guaranteed at these levels: Basic ± 5, shock ± 1, vibration ± 1, moisture resistance ± 2, load life ± 2, temperature (up to $100°C$) ± 1, What is the overall tolerance to be anticipated, assuming normal distribution for each of the characteristics noted?

7. A gear-pulley-spacer assembly consists of the following buildup:

Part	Spacer	Gear	Spacer	Pulley	Spacer
Tolerance in ±0.001	1	2	1	3	1

What overall tolerance is implied?

8. Assembly of four elements involves the basic sizes and standard deviations shown below:

	Element			
	No. 1	No. 2	No. 3	No. 4
		Millimeters × 0.001		
a. Basic size	2,000	1,000	2,000	3,000
Standard deviation	2	1.8	2	2.2

What is (A) the standard deviation of the assembly and (B) the tolerance maintainable for this assembly?

CHAPTER 15

Multiple Tolerance Chains

The methods established for components in a simple series will also work for more complex tolerance chains, involving multiple dependencies.

Tolerances for Fits

An interference fit between the hole in a gear hub and a shaft involved the dimensions and tolerances shown on lines a and b of Example 15-1. The succeeding lines show the determination of the expected tolerance for the fit, using the formula shown in line e.

Note that the value of 0.0011 inch in the third column represents the fit which occurs when the interfering extremes happen to meet; in particular, a minimum bore diameter assembled with a maximum diameter shaft. However, this is a most unlikely mating under random assembly in mass production. The probable fit for the vast majority of assemblies is defined by the average difference of 5.5 plus and minus the fit tolerance of 4.0, thus giving a range of 1.5 to 9.5 for the expected fit under random assembly.

In practical manufacture, the tolerances for the two parts can thus be opened up, and an effective fit of 0 to 11 would still be maintained. However, the previously stated conditions of statistical control and random assembly must be satisfied. Furthermore, it is important to note that the fit tolerance was computed with regard to individual parts centered at least approximately in the middle of the specification limits. In actual production, this means that the shop must "shoot for the middle" of the part dimensions. The need for such practice becomes apparent wherever processing equipment is used which is just capable or slightly better than required to meet specified limits. Generally, the variability of the items produced from such a process forms the type of distribution pattern described in curve A of Fig. 8-2, and a shift in the average will result in off-standard output.

Example 15-1.   Fit Tolerance Resulting from Assembly
of Hole and Shaft

	Hole	Shaft	Interfer-ence Fit
	(Ten-Thousandths of an Inch)		
a. Design Size*	4,510	4,521	11
b. Tolerance +	7	0	
−	0	4	
c. Average, with Product Centered Between Limits	4.513.5	4,519	5.5
d. Tolerance from Average ±	3.5	2	
e. Fit Tolerance $= \sqrt{3.5^2 + 2^2}$			±4.0†
f. Expected Fit 5.5 ± 4.0			1.5 to 9.5

*The design size of a part is that size from which the limits of size are derived by the application of tolerances. If there is no allowance for fit (clearance or interference), then the design size is the same as the basic size.

†Rounded out. This tolerance can be found directly from Fig. 12-3 by entering the graph at 3.5 and 2.0 at the left- and right-hand scales, respectively, and reading the answer of 4.0 from the center scale.

In those instances where the process variability is relatively small, so that the natural tolerance is markedly narrower than the specified limits, it is no longer desirable to aim production toward the center of the limits. An appropriate shift toward the desirable direction will then usually yield better fits and improved quality.

Tolerances and Clearances

We have just seen how the general formula for combination of tolerances will hold not only for accumulative assemblies but also for interference fits. We shall now see how realistic tolerances can be obtained, using the tolerance formula in clearance fit problems, as illustrated in Example 15-2. The data are in terms of 0.0001 inch; and the expected minimum clearance of 6.8 or 7, rounded out, is larger by 2 than the allowable minimum clearance of 5. This suggests that if the mean diameter of the hole were to decrease by 1, with an increase of 1 for the shaft, thus using up the approximate difference of 2 between nominal and expected minimum clearance, no assembly failures should result.

The effect of such a shift in dimensional averages is illustrated by the dotted frequency distribution in Fig. 15-1, which would seem to indicate at first glance that off-standard product does occur, since 2.5 percent of the holes and 2.5 percent of the shafts infringe on the

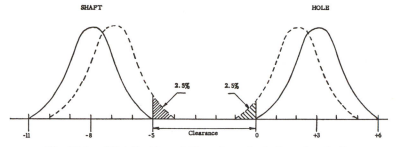

SHAFT   HOLE

2.3%   2.5%

-11   -8   -5   Clearance   0   +3   +6

Fig. 15-1.   Distribution of diameters of a hole and a shaft.

Example 15-2.   Clearance Tolerance Resulting from Assembly
of Hole and Shaft

	Hole	Shaft	Clearance Fit
	(Ten-Thousandths of an Inch)		
a. Design Size	8,000	7,995	5
b. Tolerance +	6	0	
−	0	6	
c. Average with Product Centered Between Limits	8,003	7,992	11
d. Tolerance from Average ±	3	3	
e. Clearance Tolerance* $= \sqrt{3^2 + 3^2}$			±4.2
f. Expected Clearance $= 11 \pm 4.2$			6.8 to 15.2

*Since the individual tolerances are alike, we could have used more simply $\pm 3 \times \sqrt{2}$. Here again, the nomograph in Fig. 12-3 is suitable to find the answer of 4.2.

minimum clearance specified.[1] This first impression is, however, erroneous since in actual assembly most of the oversize shafts (between −4 and −5) will usually meet with holes of +1 and wider, thus preserving the minimum clearance of 5. A similar effect may also be expected for the undersize holes. A small percentage of less than allowable clearances will of course occur, and this may be determined as shown next.

Determining Expected Failures

In illustrating the method of determining expected failures in minimum-clearance problems, we may utilize the hole and shaft

[1] The reader can readily check these percentages by referring to the normal curve (Table 6-1).

data from our last example, again in terms of 0.0001 inch. Four steps are involved:

1. *Find the Standard Deviation of the Clearance.*
   Since the clearance tolerance is 4.2, division by our (previously discussed) tolerance factor of 3 yields an estimated standard deviation of 1.4.
2. *Determine the Average Clearance.*
   With a shift in dimensional averages by +1 and −1 (7,993 and 8,002), the old average clearance of 11 has now narrowed to 9.
3. *Find the Difference Among Clearances.*
   The difference between the required minimum clearance of 5 and the average clearance of 9 is 4. Let us call this difference t.
4. *Look Up the Normal Curve Percentage.*
   The ratio $t/\sigma = z = 4/1.4 = 2.9$ is now looked up in Table 6-1 of the normal curve. We find that fewer than 0.3 percent of the assemblies will fail.

The percentage of cases in which minimum clearances will not be fulfilled is thus negligible for most practical purposes. Similarly, it can be shown that the maximum clearance limitation of 17 can be met, with practically no assembly failures.

Effect of Change in Clearance on Rejects

Let us assume that as a result of field experience with the hole and shaft assembly it is found desirable to reduce the maximum clearance from 17 to 13 while still maintaining the minimum clearance of 5 and the standard deviation of 1.4. How many rejects would be found among assemblies that were already completed?

Prior to the shift in averages just discussed in Example 15-2, we would have expected the following failures with regard to the new maximum clearance limitation:

$$\frac{t}{\sigma} = \frac{13 - 11}{1.4} = 1.43 = 8\% \text{ Assembly Failures}$$

Subsequent to the shift in averages, we obtain

$$\frac{t}{\sigma} = \frac{13 - 9}{1.4} = 2.9 = 0.1 \text{ to } 0.5\% \text{ Assembly Failures}$$

It is thus evident that, for minimum and maximum limits of 5 and 13, the design specifications based on the closer distribution averages of 8,002 for the hole and 7,993 for the shaft would be preferable for any new assemblies. In particular, they avoid the 8 percent assembly failures just demonstrated for the wider averages.

Nomograph of Expected Assembly Failures

Expected assembly failures, such as just discussed, may be determined directly by means of the convenient nomograph in Fig. 15-2 by Dr. J. N. Berrettoni (first published in Automotive Supplement No. 1 by the American Society for Quality Control). Thus, for our example of a maximum allowable clearance of 13 and an average clearance of 11, we enter the nomograph on the t-scale for t 13 — 11 or 2. Connecting this 2 with the value 4.2 on the clearance tolerance scale, we cross the diagonal line, representing expected assembly failures at 8 percent. Thus, we would expect 8 percent of the assemblies to have clearances in excess of the maximum allowed by the tolerance.

As a further illustration, for a minimum clearance of 9, an average clearance of 11, with a resultant t of 2 (as in the case above), and again with a clearance tolerance of 4.2, we would again find 8 percent—this time representing the assemblies with clearances narrower than allowed by the specified minimum of 9. It is obvious that where the minimum clearance is 0, the nomograph will yield the expected percentage of interferences.

The nomograph is thus useful in showing expected percentages which will exceed the maximum allowable clearance or fall below the minimum, as well as the expected interferences in assembly. The graph is based on the calculations shown in the preceding section.

Combination of Accumulative and Clearance Tolerances

We are now ready to examine a case history involving both accumulative and clearance tolerances, in Example 15-3, supplemented by Fig. 15-3.

Lines a to d show the conventional determination of the expected accumulation of tolerances for a bearing and snap ring, and the resultant extremes of minimum and maximum clearance within a seat. The statistical calculations, in steps e to h, show that the actual clearances may be expected to range from 3.6 to 11.4, as compared to the conventional range of 1 to 14.

Now, it was found that it would be extremely costly to maintain a tolerance of ±1.5 on the seat dimension. From the statistical analysis, consequently, it was proposed to shift the average seat dimension from 854.5 to 853 and open up the tolerance to ±3. As

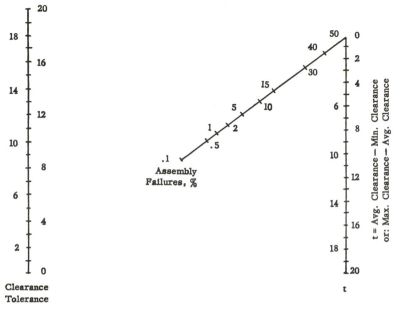

Fig. 15-2.   Expected assembly failures.

shown in lines i to l, the effect of this change is to maintain a practical range of clearances of 1.3 to 10.7, which represents a generally superior fit at a wider and therefore less costly seat dimension tolerance.

Here again, the old-timer's objections had to be answered. For obviously, if the extremes meet, there will be interference of metal. For this purpose, the old-timer agreed that one might assume that the individual parts could be classified into groups, based on the individual tolerance values, as follows:

Average				
Minus	Plus	Bearing	Snap Ring	Seat
3	. . .	724	. . .	850
2	. . .	725	118	851
1	. . .	726	119	852
0	0	727	120	853
. . .	1	728	121	854
. . .	2	729	122	855
. . .	3	730	. . .	856

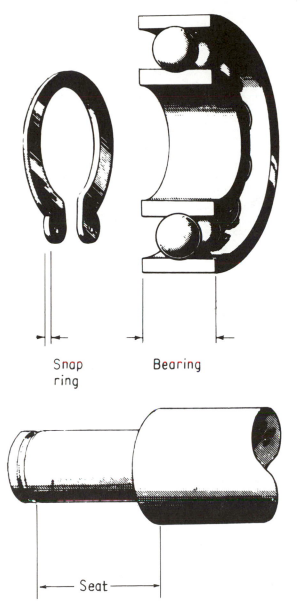

Snap
ring

Bearing

Seat

Fig. 15-3.　Assembly of bearing, snap ring, and seat. Optimal clearance tolerances
are obtained statistically, as described in Example 15-3 of the text.

Example 15-3.   Clearance Tolerance Resulting from Assembly
of Bearing, Snap Ring, and Seat

	Bearing	Snap Ring	Bearing + Ring	Seat Width	Clearance
			(Thousandths of an Inch)		
a. Basic Size	730	120		853	
b. Tolerance +	0	2		3	
−	6	2		0	
c. Maximum	730	122	852	856	
Minimum	724	118	842	853	
d. Mating of Extremes			852	853	1
			842	856	14
e. Average with Product Centered Between Limits	727	120	847	854.5	7.5
f. Tolerance from Average ±	3	2		1.5	
g. Clearance Tolerance $= \sqrt{3^2 + 2^2 + 1.5^2}$					±3.9
h. Expected Clearance $= 7.5 \pm 3.9$					3.6 to 11.4
i. New Average for Seat	727	120	847	853	6
j. Widened Seat Tolerance	3	2		3	
k. New Clearance Tolerance $= \sqrt{3^2 + 2^2 + 3^2}$					±4.7
l. Expected Clearance $= 6 \pm 4.7$					1.3 to 10.7

Since there are seven bearing dimensions, seven seat dimensions, and five ring dimensions, the total of possible combinations in assembly is $7 \times 7 \times 5$ or 245. Yet, only four of these combinations would yield interference. These are:

Bearing	Snap Ring	Bearing + Ring	Seat
729	122	851	850
730	121	851	850
730	122	852	850
730	122	852	851

These four represent just slightly more than $1\frac{1}{2}$ percent of the 245 combinations and would occur very rarely in practice, since the number of extreme dimensions in each product distribution is only a small fraction of the total product.

The nomograph in Fig. 15-2, which takes into account the fact that the proportion of extremes in each normal distribution is very small, and that each distribution has really a much finer breakdown of dimensional groups than in the simplified tabulation just given, shows that for the example at hand no interferences may be expected from a practical viewpoint.

Tolerances for Complex Combinations

The assemblies presented so far have involved simple linear combinations. Sometimes, however, tolerances must be considered resulting from assemblies involving products or quotients.

An illustration of tolerances derived from assemblies involving multiplication appears in Example 15-4. Here, output voltage of a circuit is the cross-product of the input voltage and the effects of the transformer and amplifier. With the exception of line c, the steps are self-explanatory by following the basic pattern described in previous examples. In line c, a multiplication is performed using the figures from line a. While this step actually is a procedure from calculus for obtaining so-called "partial derivatives," this need not bother the nonmathematical reader. All he needs to do is to substitute his own figures from line a in the simple multiplication in line c.

Example 15-4.   Tolerance for Cross-Product Effects, Illustration from Electronic Assembly

	Input Voltage	Transformer Ratio (Step-Down)	Amplifier Ratio	Output Voltage
a. Nominal	50	0.5	4	100*
b. Tolerance ±	1.0	0.005	0.08	
c. Derivative with Respect to Nominal Value	0.5×4 $= 2$	50×4 $= 200$	0.5×50 $= 25$	
d. $b \times c$	2	1	2	
e. $(d)^2$	4	1	4	9†
f. Output Tolerance $= \sqrt{9}$				±3

*Found from the cross-product of all entries in line a. Thus, $50 \times 0.5 \times 4 = 100$.
†The sum of the entries in line e.

TECHNICAL NOTE

Readers interested in this detail will observe that two rules of calculus were applied. The first of these says that for $u = cv^n$ the derivative $u' = ncv^{n-1}$. The second says that for a product of variables, such as $u = vwz$, the partial derivative of u with respect to, say w, is found by holding all other variables (temporarily) constant and then applying the first rule. The partials with respect to w, v, and z are thus successively vz, wz, and wv. All that remains is to substitute the input voltage of 50, the transformer ratio of 0.5, and the amplifier rate of 4 for v, w, and z, respectively.

Example 15-5.   Tolerance for Quotients, Illustrated from
Chemical Finishing Operation, Dyeing of Plastics

	Dye in Dye Bath Percent	Dye Absorbed by Product Percent	Efficiency of Absorption Percent
a. Average	4.0	3.0	75*
b. Tolerance	±0.1	±0.2	
c. Derivative with			
Respect to Average	$-100\,(3/4^2)$	$100\,(\frac{1}{4})$	
=	18.75	= 25	
d. $b \times c$	±1.875	±5.0	
e. d^2	3.516	25.0	28.5†
f. Tolerance $= \sqrt{28.5}$			±5.3

*Determined from $3.0 \div 4.0$ expressed as a percent.
†The sum of the entries in line e.

TECHNICAL NOTE
    Even though we deal with a quotient, $v/w = 3/4$, we can use the rule for products by rewriting $v/w = vw^{-1}$, where $w^{-1} = 1/w$. With respect to v, the partial derivative of $vw^{-1} = 1/w$, since any number raised to the zero power is one. With respect to w, the partial becomes $-vw^{-2} = -v/w^2$. All that remains is to substitute $v = 3$ and $w = 4$, and then to revert back to percentages by multiplying by 100.

Tolerances involving quotients are evaluated by steps parallel to those for products. An illustration from a chemical finishing operation is given in Example 15-5, in which the reader can again substitute his own data to obtain the answer sought.

Evaluation of Mechanism Reliability

A special aspect of tolerances, which has gained prominence particularly in the electronics and missile industries, is reliability evaluation. Here, reliability usually refers to the likelihood of nonfailure in the performance of an item. Thus, if a part has a reliability of 0.99 or 99 percent, this means that past experience with the particular product indicates that it will function properly 99 times out of 100, with one failure per 100.

Problems of reliability are especially urgent in complex assemblies, consisting of large numbers of intricate parts. Thus, if a mechanism has 400 parts, each of which has a reliability of 99 percent, then the assembled mechanism will have a reliability of $(0.99)^{400}$, which comes out to an unacceptably poor 2 percent. (The statistical basis of this type of calculation was demonstrated in Chapter 3.) No one would, in general, accept a mechanism with a 98 percent expectation of failure. Manufacturing methods must be such, therefore, that most of the parts have a reliability which for all practical purposes is 1.0, with only a few items of lesser reliability. Thus, the reliability of the entire mechanism may be:

$$\text{Mechanism Reliability} = 1.00^{397} \times .99^1 \times .98^2 = 95\%$$

Much of the work in reliability engineering consists of isolating the type of parts which have relatively low reliability, and seeking corrective means to bring mechanism reliability up to the required standard.

Computer Simulation

In those cases where assemblies are relatively complicated, such as in the case of eccentric locations and mounts, the statistical formula for tolerance evaluation may become quite cumbersome. It is then preferable to feed the actual frequency distributions of the component parts into an electronic computer, then let the machine combine the parts data at random, and write out the resultant frequency distribution of the assembly.

A typical problem of this nature is illustrated in Fig. 15-4 for the frequency distribution (A) obtained by random combination on a computer of the patterns of eccentricities (B) and (C). In imitating or "simulating" random assembly on the computer, a set of random angles and cosines was incorporated in the program. The resultant distribution obtained represents the pattern which may be expected

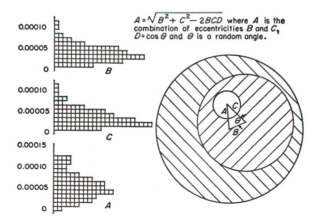

Fig. 15-4.   Frequency distribution A, obtained from eccentricities B and C, for an eccentric shaft supported by an eccentric mount.

in actual assembly. This illustration is courtesy of Mr. R. L. Thoen, staff engineer, General Mills Incorporated, and of *Machine Design* who kindly granted permission for its use.

Special Tolerancing Problems

Despite the wide use and applicability of the general statistical tolerancing formula, there are some exceptions. These occur primarily when distribution patterns are lopsided or "skewed." When skewness occurs, the distribution does not exhibit symmetry, and a "tailing out" toward either extreme side occurs. For example, all three distribution patterns in Fig. 15-4 show marked skewness toward the high side. The general analysis of variance formula presented will then be inapplicable, and statistical analysis in general will become quite complex. We therefore have only one of two choices:

1. Allow for tolerances of assemblies by means of simple (nonstatistical) additive procedures.
2. Seek to predict the probable distribution pattern of the assembly, by means of the computer simulation discussed above.

Where access to a computer can be had, the second method will generally be found to be relatively inexpensive in relation to the benefits obtainable from tolerances that are as realistic as feasible.

The problems presented by skewed distributions are the reason why the reader has been advised to check frequency patterns, obtained from routine control chart data. In many instances, such as production involving eccentricities, radial play, screened and sorted product, or small lots, a check of distribution patterns is especially advisable. In other cases, where experience has shown that skewness need not be expected, a check of frequency patterns may be deemed superfluous. Often, of course, a check of frequency distributions will reveal skewness where none ought to be, such as for example in certain dimensions of ball bearings, thus indicating where production needs to be corrected.

The task of reaching optimality in tolerancing is thus part of an effective overall quality control program for the plant. Ultimate costs are greatly affected by the care which has gone into establishment of proper specifications and tolerances for the components of mass production. As pointed out by Earle Buckingham: "There is probably no other place in the organization where so much money can be saved by careful attention to detail, and there is certainly no other place where so much money can be wasted by carelessness and ignorance."[2]

Dynamic Programming

The determination of realistic tolerance chains can be carried a step further, by considering the interplay among various cost alternatives. Optimum tolerancing, then, does not only consider statistical aspects but also identifies that combination which involves lowest total costs, consistent with overall specification requirements. The method used for this purpose is known as a part of the general technique of *dynamic programming.*

We shall present a simple illustration, using as our basis Example 15-3, where there is a clearance tolerance of ± 4.7 in the assembly of bearing, snap ring, and seat. For each of these three parts, a tolerance of ± 2 or a tolerance of ± 3 can be produced, but at different costs. For the bearing and snap ring, the costs are as shown below:

Example 15-6.   Tolerances and Costs in Bearing and
Snap Ring Assembly
(First Phase of Dynamic Programming Analysis)*

			Bearing		
	Tolerance	± 2	Tolerance		± 3
	Squared Tolerance	± 4	Squared Tolerance		± 9
	Cost	$2.50	Cost		$1.50
All Dimensional Data in Ten-Thousandths of an Inch					
	Tolerance	± 2			
	Squared Tolerance	± 4	± 8		± 13
Snap Ring	Cost	$1.80	$4.30		$3.30
	Tolerance	± 3			
	Squared Tolerance	± 9	± 13		± 18
	Cost	$1.30	($3.80)†		$2.80

*Entries in each cell show the effect of combining squared tolerances and costs.
†This $3.80 cost deserves no further consideration, since another assembly of equal tolerance (see upper right-hand cell) costs less ($3.30).

[2] *Dimensions and Tolerances for Mass Production,* The Industrial Press, New York, N.Y., 10016, page 17.

The entries on the upper and left-hand margins will be self-explanatory. In the four cells, the effect of combining squared tolerances (recall that one cannot add simple tolerances!) and costs is given. We might have included a wider range of tolerances, such as ±1.5, ±2, ±2.5 ± 3, or further breakdowns. Such would have been more "realistic" but would mar the clarity of presentation, and the principal points to be made.

We have completed the first phase of analysis. A second phase is now needed, to include consideration of the seat tolerance. For this purpose, we transfer the applicable cell values of the prior compilation to the top section of the presentation below. Next we add the seat dimension tolerances and cost factors.

Example 15-7.   Tolerances and Costs for Bearing, Snap Ring
and Seat Assembly
(Second Phase of Dynamic Programming Analysis)

Bearing Plus Snap Ring Together

	Tolerance Squared	±8	Tolerance Squared	±13	Tolerance Squared	±18
	Cost	$4.30	Cost	$3.30	Cost	$2.80

All Dimensional Data in Ten-Thousandths of an Inch

	Tolerance	±2			
	Squared Tolerance ±4		±12	±17	±22
Seat	Cost	$1.20	$5.50	$4.50	$4.00
	Tolerance	±3			
	Squared Tolerance ±9		±17	±22	(±27)†
	Cost	$1.00	$5.30	$4.30	($3.80)

NOTE: Entries in each cell show the effect of combining squared tolerances and costs. Lowest cost occurs at $4.00 for the combination in upper right-hand corner. This corresponds to a bearing of ±3, a snap ring of ±3 and a seat of ±2 as regards tolerance.

†This squared tolerance of ±27 is not permissible, since the overall tolerance is specified at ±4.7, and squaring this value yields a maximum allowable overall of 22 (rounded).

It is simpler to find out from this tabulation what happened than to talk about it. The lowest-cost and yet acceptable tolerance is obtained when we combine a bearing of ±3 tolerance with a snap ring of ±3 tolerance and a seat of ±2 tolerance. The total cost is

$1.50 + $1.30 + $1.20 or $4.00 even. Note again that we do not claim that it is sufficient to investigate the limited number of tolerances we checked, nor is there any claim that the cost values are generally applicable. The emphasis is on the method and the principles of analysis. Once these are understood, more refined and complex analyses follow readily. Moreover, computer programs under the general heading of "dynamic programming" are available for the performance of calculations. In order to apply such "canned programs," however, the basic aspects of the methodology involved should be recognized by the user.

A word about the applicability of the term "dynamic programming" is in order. "Dynamic" implies a multiphase situation. In the present instance, we have used two phases, but for more complex tolerance chains a multitude of phases may be involved. "Programming" does not imply "computer programming," but rather it refers to a program of carefully scheduled phase-by-phase analyses of a problem.

We will encounter dynamic programming again, in connection with the optimization of reliability designs, in Chapters 16 through 18 of this book. The multiple ways in which an analytical technique can be applied to a variety of problem situations is thus underscored.

Practical Tolerancing

In the two chapters now coming to a close, we have discussed the essentials of statistical tolerancing. The methods are sound and workable; yet there are often failures in the installation of this approach. The principal technical reason for such problems, it seems, lies in weaknesses in the integration of the tolerance system. Yet, it is only within a systems context that statistical tolerancing will work. The following factors must be meshed:

1. Tolerances for the final product must recognize customer and market needs, giving due regard to the elements of quality, cost, and production time.
2. Component tolerances should form a chain that achieves the assembly tolerance adequately, but at minimal cost.
3. Production processes should be set so as to produce the requisite parts tolerances.
4. Important dimensional and other quality characteristics require surveillance by means of control charts.

5. Analysis of variance studies should assure maintenance of optimal uniformity of product, consistent with cost and time factors.
6. Product design, review, and redesign must become an integral part of the tolerancing cycle.

For proper coordination of all of these elements, it is essential that the persons concerned with the tolerancing program understand the statistical, economic, and technological reasoning involved. Moreover, an attitude of acceptance, cooperation, and even enthusiasm must be developed. It is the presence of both—sound technological and engineering foundations and effective human relations—which together serve the purpose of attaining the benefits of statistical tolerancing.

Broadened Scope of Tolerances

The statistical methods developed in this chapter can and have been expanded to cover all areas of managerial decision processes. Example 15-8, which is concerned with the investment evaluation of risks and profits in a new product, should involve persons from sales, accounting and finance, and production, including quality control. Using the results of market studies, sales forecasts, pricing analyses and cost projections, the group applies judgment to come up with an average estimate (line a) as well as the range between the most optimistic and most pessimistic opinions (lines b, c, and d). Next, the individual standard deviations (line 3) lead to the overall standard deviation (line h) of $15,000 by steps running parallel to those in Examples 15-4 and 15-5. Noting an average expected profit of $30,000, what is the risk now of a break-even ($0 profit) or loss? Using the method of Chapter 6, form the z-ratio of ($30,000 $-$ 0)/$15,000 which equals 2. From the normal curve Table 6-1, at the level 2, the risk value is 2.3 percent.

Various competitive proposals may be evaluated and ranked by means of this approach, contrasting potential gains versus risk. The final go-ahead is given to those projects which seem most worthwhile. Quality control's involvement is based on the need to assess control capabilities for the new product as regards costs and quality, which in turn affects anticipated profits and salability. This role is consistent with the concept of a total quality control program, as brought out in Chapter 10 on organizational aspects.

Example 15-8.   Statistical Tolerancing Applied to New Product Investment Decision

Estimates and Calculations	Sales Quantity, Q, Units	Profit Margin, M, $/Unit	Fixed Cost, F, $	Total Profit, P, $
a. Group Average	20,000	10	170,000	30,000*
b. Most Optimistic	21,000	11	155,000	76,000
c. Most Pessimistic	15,000	8	185,000	−65,000
d. Maximum Range = b-c or c-b	6,000	3	30,000	
e. Individual Standard Deviations = d/6**	1,000	0.5	5,000	
f. Partial Derivative	10	20,000	1	
g. $(e \times f)^2$ in 1,000,000	100	100	25	225
h. Standard Deviation, Overall = $\sqrt{225,000,000}$				15,000

---

*$P = QM - F = \$20,000 \times 10 - \$170,000 = \$30,000$. Proceed similarly for lines b and c.
†Divisor 6 based on normal curve, $(\pm 3 = 6)$.
NOTE: Lines f through h follow the rules of Examples 15-4 and 15-5.

REVIEW QUESTIONS

1. Does the formula for linear combinations in assembly work also for fit tolerances? (Give reasons for your answer.)
2. Does the formula just mentioned work also for clearance tolerances? (Give reasons.)
3. How is it possible to predict assembly failures for various combinations of diameters, such as for a hole and a shaft, where a definite clearance tolerance is required?
4. In question 3 what are the specific assumptions that must be fulfilled if such predictions are to be successful?
5. A mechanism consists of three components, each of which has an 80-percent reliability. What is the overall mechanism reliability?
6. Under what circumstances would one wish to resort to computer simulation in evaluating tolerance combinations?
7. Why should one wish to predict tolerancing effects in advance of production and assembly? Why not proceed "from practical experience" and then make corrections as we go along?
8. What is the role of dynamic programming in tolerancing work?
9. Tolerances in 0.0001 inch involving a hole and shaft assembly are as follows:

Components	Hole	Shaft	Clearance
Design size	4,000	3,990	10
Tolerances	+8	−8	

What are (A) clearance tolerance and (B) anticipated clearance?

10. A clearance tolerance is ± 0.0003, while the average clearance is 0.0010. What percent of the product will fail to meet the minimum clearance of ± 0.0007 inch?

11. In the construction of roads and airfields, the plastic index is an important concept. A soil with an index greater than 15 is not suitable until it has been properly treated. Plastic index is from laboratory tests on soil samples by subtracting the liquid limit value from the plastic limit value. The following are typical results:

	Average	Standard Deviation
Liquid limit	40	3
Plastic limit	25	4

What is the standard deviation for plastic index? What tolerance should be attainable under the conditions stated?

12. Base materials, also, must be tested for proper plastic index. Typical averages for liquid and plastic limit are 20 and 10 with standard deviations of 2 and 3. Again, what is (A) the standard deviation for the plastic index and (B) the tolerance (±) attainable?

13. A lamination consists of five layers with the average dimensions as shown below.

Layer:	Metalized mylar	Adhesive	Vinyl Film	Adhesive	Backing Paper
Dimension, 0.001 inch:	2	1	3	1	2

What individual tolerances are required, assuming identical limits for each layer, if the overall tolerance is to be within ±0.0005 inch?

14. Assembly of a hole and a shaft involved the basic data listed below in 0.001 millimeters.

	Hole	Shaft	Interference Fit
	millimeters x 0.001		
a. Design size	2,990	3,000	10
b. Tolerance	+6	0	
	0	−4	

What is (A) the fit tolerance and (B) the expected fit?

See Cases 9 through 11, Appendix 3, for additional review materials.

CHAPTER 16

Evaluation of Reliability

"The most nerve-wracking part of any space flight is the fact that your life depends on thousands of critical parts, each produced by the lowest bidder."

This comment, quipped by an astronaut, is a nutshell evaluation of an intense quality assurance problem: How to enhance reliability of equipment by minimizing the possibility of malfunctions. It is not just in space vehicles, however, that reliability is essential. For example, a large-scale computer consisting of many thousands of critical parts, cannot be permitted to have undue breakdowns. Yet, with growing complexity and increasing numbers of critical parts, electronic computers, automation machinery, and many consumer products, are becoming more and more subject to malfunction problems. There is a great need for thorough and effective quality con- trol in the production of such equipment. Reliability assurance, however, goes a step beyond quality control by providing engineer- ing proof of the performance capability of products.

The problems of reliability in complex systems, such as missiles and electronic equipment, have been mentioned briefly on page 210, Chapter 15. We may restate in more precise terms the nature of reliability as the probability that *a product, device, or equipment will give failure-free performance of its intended functions for the required duration of time.* While reliability as such is not new, the concept of assessing it in quantitative terms with a conscious engi- neering effort toward reliability improvement has been with us for only a short span of years.

Liabilities incurred as a result of unreliability may be enormous, and in fact law suits arising from allegedly unsafe products as regards design, manufacturing, quality control, and quality assur- ance have been increasing rapidly. A well-known instance is the action by the widow of Lieut. Col. Virgil I. Grissom for the death of her husband in a launching-pad fire, filed against the builders of the Apollo spacecraft. A board of review, set up by the Federal space agency, had revealed what the suit described as "many deficiencies in design, engineering, manufacture, and quality control." In addition, poor installation and workmanship were cited. It seems certain that a firm without effective quality controls, reliability assurance, and careful documentation of tests and inspections will be unable to effectively defend a damage suit.[1]

Types of Failure

Sound evaluation of reliability begins with a consideration of the types of failure that may be encountered. Engineers generally have defined three types of failure:

Early Failures. These are the failures resulting from production defects or other deficiencies. Usually, they will show up relatively early in the life of a product. The practice of "debugging," "shaking-out" and "burning-in" on new products and equipment is designed to discover and correct such deficiencies.

"Chance" Failures. These are failures that may occur at various intervals during the life of an equipment. Hidden defects that escaped the "early failure" period may result in malfunction later. Environmental stresses, such as electrical, magnetic, temperature, or vibration effects, may at times interact in such a manner as to cause component parts and thus the system as a whole to fail. The term "chance" refers to the fact that a particular failure is relatively unexpected. The rates at which such failures are likely to occur, however, have been studied for many components, subsystems, and equipments as a whole. Fortunately, these rates are quite low in the vast majority of components. Otherwise our modern complex technology would be impossible.

People sometimes say that there cannot be any such thing as a "chance" failure, because every malfunction or breakdown has a cause. True enough, there is such a cause. But it is often extremely difficult to trace it, especially when failures occur at such rare intervals as once every 100,000 or million hours. Yet, we are concerned about these failure rates, since failure of one among many thou-

---

[1]Legal aspects of quality control and reliability, practically unheard of a decade ago, are consuming a fair amount of space in the technical literature. A recent book by George Peters, *Product Liability and Safety*, Washington, D.C.: Coiner Publications, 1971, covers this topic.

sands of components in a complex system may bring about system failure. We continue to seek means of reducing chance failure rates to the extent practical, feasible, and economical.

Wear-Out Failures. This is probably the most inevitable of all types of failures. Examples abound. Abrasion may cause piston rings, cylinders, valves, and bearings to break or function unsatisfactorily. Fatigue or creep of metal, chemical decomposition and corrosion, radiation damage and vibration all take their toll on product life. Wear-out failure occurs when the effect of wear interferes with the intended applications of an object.

For a typical cycle of failures, the fate of humans is often given as a telling example. In the very early stages of life, birth defects cause a relatively high mortality. Then for many decades, the death rate remains relatively low and stable, with only occasional "chance" failures from accidents or relatively rare illnesses or diseases. Then, as we approach later years, the "wear-out" stage takes its toll.

Let us examine an automobile tire. Early failures may represent major defects in production, resulting in rejection before sale or else in unsatisfactory performance during the first few days or weeks of use. Thereafter, a tire is likely to run trouble-free for many thousands of miles, excepting that there remains a small probability of failures: road hazards and driving hazards may result in a blow-out, a valve may begin to leak air, or a slow leak may develop elsewhere. Corrective maintenance, such as replacing the leaky valve and preventive maintenance, such as care to properly inflate and rotate tires at periodic intervals, can help prolong the life of the tire. Finally, however, as the tire wears bald, the battle against punctures and possible blow-outs becomes a losing one, and it is wise to make a replacement.

Inasmuch as wear-out failures are most readily understood from experience, their quantitative statistical nature will be examined first. Next, chance failures will be considered. Early failures will not receive extensive attention here. The testing and quality control program of a company will remove such defective products through final inspection, "debugging," "shaking-out," or "burning-in" procedures.

Analysis of Wear-out Failures

In order to examine the statistical nature of wear-out failures, we may utilize an example of product life testing on electron tubes.

A production lot of electron tubes, especially designed to withstand high temperatures and intense vibrations, has just come off the line. A sample of 25 tubes is placed in a tray and subjected to life testing under the specified heat and vibration conditions. As each tube fails, an automatic recorder shows the time of failure. When the last of the 25 tubes has failed, the data shown on page 223 in columns (1) to (6) are noted below. From these, it is a simple matter to derive the cumulative failure and survival percentages, as well as the force-of-failure data of columns (7) to (10), supplemented by the graphic presentation of Fig. 16-1.

If the test was realistic in terms of the required temperatures, vibrations, and other ambient conditions, then it will have yielded useful information regarding the survival characteristics of the tubes. There will be justification for a statement, based on the percentage of survivals of column (9), as follows:

"Under actual operation, we may expect that the tubes will survive well enough so that:

92 percent will operate properly for at least 109.9 hours,

76 percent will operate properly for at least 119.9 hours, and

28 percent will operate properly for at least 129.9 hours, but beyond that only a few tubes will function."

Whether or not the production lot conforms to specifications can

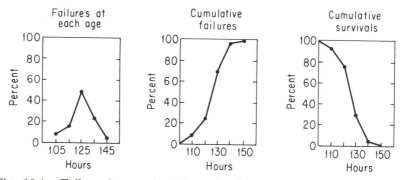

Fig. 16-1. Failure frequencies. First, the failure frequency at each tube age is shown. Next, these frequencies are cumulated upward, showing the percentage of tubes that failed after 110, 120, etc., hours or at a later age. Finally, a downward cumulation of failures gives, in effect, the percentage of tubes that survived at least up to the ages 110, 120, etc., hours of the base scale. The percentages of survival also represent the probability that an individual tube chosen randomly from the lot will survive to at least 110, 120, etc., hours.

Table 16-1.   Analysis of Wear-out Failures

(1) (2) Time Interval Hours in which Tubes Failed		(3) (4) Number of Good Tubes at		(5) (6) Failures during Interval		(7) (8) Cumulative Failures		(9) Cumulative Survivals	(10) Force of Failure
Start	End	Start	End	No.	%	No.	%	%	%
100	109.9	25	23	2	8	2	8	92	8
110	119.9	23	19	4	16	6	24	76	17
120	129.9	19	7	12	48	18	72	28	63
130	139.9	7	1	6	24	24	96	4	86
140	150	1	0	1	4	25	100	0	100
Totals				25	100				

NOTES:

Columns (1) and (2) give the time interval in which the various tubes failed.

Columns (3) and (4) show that we started with a sample of 25 tubes. As tubes failed during each time interval of 10 hours, the number of "good" tubes surviving dwindled. Except in the first line, the start of Col. (3) is the end of Col. (4), prior line.

Columns (5) and (6) contain the number and percent of tubes failing in each time interval. The original 25 good tubes serve as the percentage base. For the second row as an example, 4 failures out of 25 tubes = $100 \times 4 \div 25 = 16$ percent.

Column (7) is cumulated by successive additions from Col. (5). For example: $0 + 2 = 2$, $2 + 4 = 6$, $6 + 12 = 18$, etc. Col. (8) is obtained similarly from Col. (6).

Column (9) represents a subtraction of Col. (8) from 100 percent.

Column (10) reveals an increasing force of failure, based on the relation:

$$\text{Force of Failure} = 100 \times \frac{\text{No. of Failures During Interval}}{\text{No. at Start of Interval}} = 100 \times \text{Col. (5)} \div \text{Col. (3)}.$$

For example, for the second line, $100 \times 4 \div 23 = 17$ percent (rounded).

then be evaluated in terms of the intended use of the tubes. We shall examine these problems further, beginning with a consideration of how the cumulative survival percentages lead to probability and thus reliability statements.

Probability of Survival

We have observed that 92 percent of the tubes in the sample survived for at least 110 operating hours. From this, it is our most likely expectation that this same percentage in the production lot will also survive for at least 110 hours. Moreover, it is intuitively clear that for any one tube chosen at random from the lot, the probability of survival for at least 110 hours is 92 percent. We have moved from a relative frequency to a probability statement. This approach is also consistent with the probability concepts developed in Chapter 3 and subsequent materials in this book.

True enough, we cannot say whether a given tube will have a certain length of age. In fact, a tube may survive anywhere from 100 to 150 hours. But the chances that any tube will survive at least 110 hours are 92 out of 100. Correspondingly, the risk of failure is 100 — 92, or 8 percent.

Reliability in terms of probability of survival is often denoted by the expression P_s given in terms of a decimal fraction or a percentage. Thus, P_s for at least 110 hours is 92 percent or 0.92.

Sharpening the Reliability Assessment

We have just completed a valid assessment of product reliability. Our evaluation can be sharpened, however, if we can make use of the underlying distribution of the lot for a further, somewhat more refined estimate. The normal distribution pattern of the illustrative example of the electron tubes justifies our assuming a normal pattern for the lot itself.

There is still another reason for presuming a normal lot distribution. Experience for a large variety of products in many different industries shows that most often the observed patterns of wear-out tend to be normally distributed about the arithmetic mean of the wear-out age. Accordingly, by assuming on the basis of the available evidence that the wear-out phenomena for our electronic tube lot are normally distributed, we can reconstruct this distribution by reference to the average failure age of 125 hours and the computed standard deviation. For the latter quantity, we use the sampled

Table 16-2.   Standard Deviation of Tube Wear-out Data

Midpoint Failure-Time Interval*	Frequency of Failures†	Deviation from Mean‡	Squared Deviation	Frequency × Deviation‡
105	2	−20	+400	800
115	4	−10	100	400
125	12	0	0	0
135	6	−10	100	600
145	1	−20	400	400
Totals	25	—	—	2,200
Variance (2,200/24), in hours				92†
Standard Deviation, $\sqrt{92}$, rounded, in hours				10

*Each entry is the midpoint of the 100 to 110, 110 to 120, etc., intervals in which failures are recorded.

†These are the same frequencies as previously given.

‡Arithmetic mean is 125 hours.

§The sample size, N, is 25. Use of $N - 1 = 25 - 1$ or 24 in the divisor is a refinement of the calculation procedure, which is further explained in Appendix I.

data on age at time of failure as a basis, proceeding as shown in Table 16-2.

The standard deviation found is 10 hours. This value, together with the arithmetic mean of 125 hours, serves as the estimator of the lot characteristic. Instead of the ordinary bell-shaped normal curve, however, we shall use its cumulative form as shown in Fig. 16-2. This smooth curve is similar to the pattern revealed in the third graph of Fig. 16-1. The difference between the two is that the latter represents the sampled data as received, whereas the former is the smooth normal curve.

The vertical scale of the cumulative normal curve is expressed in terms of probability of survival, P_s, from 0 to 100. We realize that the 100-percent point more accurately represents 99.7 percent. The base scale is in terms of plus and minus deviations, in multiples of standard deviation, from the center of the distribution. The center represents the arithmetic mean of the data. Zero at the base scale of the cumulative normal curve is therefore that point which corresponds to the arithmetic mean, or 125 hours in our example. The curve shows, as expected, that the probability of survival, P_s, to at least 125 hours for any tube is 50 percent. Next, the point "−1 standard deviations" corresponds to $125 - (1 \times 10)$ or 115, inasmuch as σ is 10. We observe that P_s is 84 percent. Other points are obtained similarly.

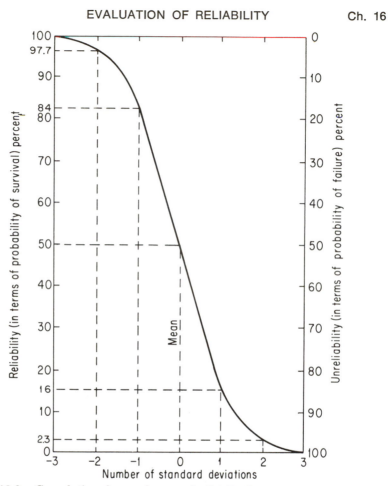

Fig. 16-2. Cumulative form of normal distribution curve, expressed in probability terms for use in reliability evaluations. Observe that the arithmetic average occurs at the 50-percent probability point.

By employing the normal curve for our estimations, we have smoothed the data and thus increased the precision of the method of evaluating reliability. Let us make a quick comparison. Previously, the cumulative data of Fig. 16-1 yielded a P_s for 110 hours' failure-free operation of 92 percent. The smoothed normal curve approach proceeds as follows:

1. The point of "at least 110 hours' survival" corresponds to 15 hours below the average of 125.

2. Show these 15 hours as —15 to indicate the "below average" position.
3. Divide —15 by the standard deviation of 10, giving —1.5.
4. In Fig. 16-2, —1.5 corresponds to 93 percent for P_s.
5. As an alternative to the graphic evaluation, refer to Table 6-1, which has the entry 43 at 1.5 σ. (Since the table deals with only one half of the curve, pluses and minuses are omitted). Add 50 to 43, giving 93 percent for P_s.

Although the two reliabilities of 92 and 93 percent for the original and smoothed data are quite close, as we reach farther points of the curve, the discrepancies tend to become larger. Hence it is preferable to use the smoothed curve.

Analysis of Chance Failures

Between the relatively brief phases of early failures and later wear-out there is a comparatively long period of useful life of a product, during which there is usually but a small expectation of malfunction. For example, a production lot of a certain item may contain some 2 percent of the items that fail during burn-in. Next, there is a useful life of, say, 10 hours with an expectation that some 0.5 percent of the items may fail, followed by the wearout stage.

Now if 10,000 units have been burned-in, they will next be entering a useful life expectation of a total of 10,000 times 10 hours or 100,000 hours. Fifty items, representing 0.5 percent of 10,000, will fail at some time in this span. These are the chance failures. In quantitative terms we may say that the failure rate, denoted by a small Greek lambda (λ), is:

$$\lambda = 50 \div 100{,}000 = 0.0005 \text{ failures per hour}$$
$$\text{or 5 failures per 10,000 hours.}$$

Conversely, one may say that the average or mean time to failure, designated by MTTF or more briefly by m, is

$$m = 100{,}000 \div 50 = 2{,}000 \text{ hours}$$

Observe that $m = 1 \div \lambda$ and $\lambda = 1 \div m$, so that $m \times \lambda$ is always 1.0 and the two expressions are the reciprocals of each other. While the illustrative example has given failure rates that seem low; in practice, rates that are many times lower will often be needed. This is especially true for complex systems consisting of many thousands of components, when proper operation of all parts is critical for failure-free function of the system as a whole. The manner in which small unreliabilities of individual parts can cascade into excessive risk of system failures was previously emphasized (see page 210, "Evaluation of Mechanism Reliability").

Distribution of Chance Failures

The occurrence characteristics of chance failures have been studied extensively on a wide variety of products. From this work, it has been found that the "normal distribution" may be expected relatively rarely. Instead, a so-called "negative exponential distribution" can be anticipated most often. A leading specialist in reliability testing sums up the results of considerable experience:[2]

> "It seems as though the exponential distribution (of chance failures) plays a role in life testing analogous to that of the normal distribution in other areas of statistics."

An exponential distribution arises whenever the force of failure remains essentially constant throughout the time period considered. The illustration below demonstrates the occurrence of chance failures in a 1,500 hours period for a 1,000-part lot, each part being a transistor. The force of failure remains constant at 10 percent (unlike the increasing force of failure in normal distributions). This percentage, expressed as a decimal, is also known as the *instantaneous failure* or *hazard rate*.

Each time interval shown in the transistor example represents 100 hours. For the 1,000 transistors at the start, there was a total operating time of $1,000 \times 100$ or 100,000 hours. The 100 failures then represent: $\lambda = 100 \div 100,000 = 0.001$ failure per hour or one failure per thousand hours. Moreover, $m = 100,000 \div 100 = 1,000$ hours.

Because the force of failure is constant at 10 percent or 0.1 for all intervals, λ and m are 0.001 and 1,000 hours, respectively for the entire set of data. Reliability in terms of P_s is given by the cumulative percentages of column (9), Table 16-4, and is graphed in Fig. 16-3. The pattern of the curve is characteristic of a "negative exponential" distribution. "Negative" refers to the downward sloping nature. The term "exponential" indicates how P_s is obtained. For this purpose, let us first recognize that a 10-percent force of failure represents also a 90-percent or 0.9 force of survival. Now, in

---

[2]Benjamin Epstein, "The Exponential Distribution" in *Industrial Quality Control*, vol. xv, no. 6 (December, 1958), page 4. Parentheses added.

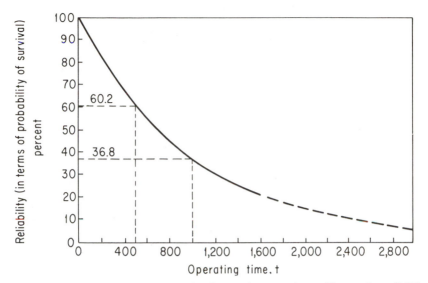

Fig. 16-3. Reliability curve, derived for the transistor illustration. Solid line represents demonstrated reliability, dotted line represents extrapolation.

the first time interval we had 100 percent survivors to start. At the end, we had 100×0.9 or 90 percent left. In the next interval we had 90 percent $\times$ 0.9 or 81 percent, and in the third interval $81 \times 0.9 = 73$ percent left. Thus, an exponent is operating, because:

$$P_s = 100 \times 0.9 \times 0.9 \times 0.9 = 100 \times 0.9^3 = 73 \text{ percent}$$

Other values of P_s are found similarly, but it becomes tedious to keep multiplying for exponential results. A simpler approach is to use a standardized exponential curve.

Standardized Exponential Curve

This curve is shown in Fig. 16-4. It has been standardized by showing, in place of actual hours, the ratio t/m, representing an actual or required operating time, t, of equipment to the mean time to failure, m. Assuming, now, a t of 300 hours, then $t/m = 300/1,000 = 0.3$.

Entering the graph at the base point 0.3, we cross the curve at P_s of 74 percent. This is 1 percent different from the 73 percent just found. The standardized curve, for reasons not easily demonstrated by elementary methods, gives results slightly more precise than obtained by the procedure of the preceding section.

Table 16-3.   Effect of Constant Force of Failure on Reliability Obtained

(1) Time Interval, Hours, in Which Parts Failed		(3) Number of Good Parts at:		(5) Failures During Interval		(7) Cumulative	(8)	(9)	(10) Force of Failure
						Failures		Survivals	
Start	End	Start	End	No.	%	No.	%	%	%
0	99.9	1000	900	100	10	100	10	90	10
100	199.9	900	810	90	9	190	19	81	10
200	299.9	810	729	81	8	271	27	73	10
300	399.9	729	656	73	7	344	34	66	10
400	499.9	656	590	66	7	410	41	59	10
500	599.9	590	531	59	6	469	47	53	10
600	699.9	531	478	53	5	522	52	48	10
700	799.9	478	430	48	5	570	57	43	10
800	899.9	430	387	43	4	613	61	39	10
900	999.9	387	348	39	4	652	65	35	10
1000	1099.9	348	313	35	3½	687	69	31	10
1100	1199.9	313	282	31	3	718	72	28	10
1200	1299.9	282	254	28	3	746	75	25	10
1300	1399.9	254	229	25	2½	771	77	23	10
1400	1500	229	206	23	2	794	79	21	10

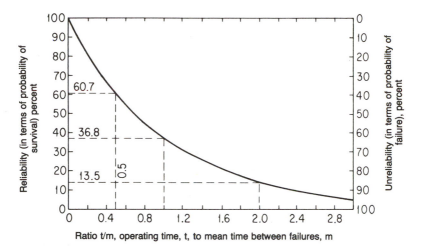

Fig. 16-4.   Reliability curve, based on the ratio of operating time, t, to mean time to failure, m; assuming an exponential distribution (constant force of failure).

The standardized exponential curve is computed from the relation

$$P_s = (1/e)^{t/m}$$

where e is the base of the natural logarithm, 2.71828+ or more simply, 2.72. Hence, $1/e = 0.368$, so that for $t/m = 0.3$, $P_s = 0.368^{0.3} = 0.74$ or 74 percent. Equivalent forms of identical meaning found in the literature are $P_s = e^{-t/m}$ or $P_s = \exp(-t/m)$. For a listing of detailed values, see *Tables of the Exponential Function, e^x*, published by the U.S. Department of Commerce.

A noteworthy result of the exponential distribution is the fact that the probability of survival, P_s, of an object to its mean time, m, is not 50 percent (as in the normal distribution) but only 36.8 percent. Survival probability to the point where $t = 2m$, or in other words the chances that an object will survive for twice as long as its average life expectation, is only 13.5 percent. These probabilities can be read from Fig. 16-4 or else gleaned from the exponential expression just given.

Table of Reliabilities

In lieu of the standardized curve, tabular values (in Table 16-4) may be preferred. For our transistor example, we had λ of 1 failure per 1,000 hours and thus 100 per 100,000 hours, corresponding to an m of 1,000. Entering the table at 100 failures per 100,000 hours, we find that we may expect a transistor to survive at least to 100, 200, 500, or 1,000 hours at these respective probabilities in percent: 90.5, 81.9, 60.7 and 36.8. The values agree closely with those obtained in the tabulated calculations previously given on page 230.

The example also permits a further check of the expression $0.368^{t/m}$. Thus, a survival time t of 500 hours yields a t/m of $500/1,000 = \frac{1}{2}$. But a number raised to the half power is its square root. The square root of 0.368 is 0.607 corresponding to the 60.7 percent above. When t is 2,000, t/m becomes 2, and the square of 0.368 is 0.135, which corresponds to the entry $P_s = 13.5$ percent under $t = 2,000$ of the table. When t is 1,000, t/m is 1, so that $0.368^1 = 0.368 = 36.8$ percent as above.

A valuable further application of the table is apparent. Suppose that we had checked the transistors for only 100 hours, observing a force of failure of 10 percent, from which m is then found to be 1,000. So long as we can reasonably expect the force of failure to be constant, we can make reliability predictions for survival of any transistor to 200, 500, 1,000, 2,000, or 3,000 hours or more, using merely the ratio t/m. The exponential curve that started out as a theoretical discussion has thus become a practical tool for evaluation and prediction of reliability.

Repairable and Nonrepairable Failures

The concept of mean time to failure (MTTF), or m for short, has been used with regard to equipment that for economic or practical reasons is not repairable, such as an orbiting satellite. When repairs are expected, however, we use mean time between failures (MTBF). Both terms are designated by the shorter symbol, m, since they are used in identical manner in entering Fig. 16-4 or Table 16-4.

Many objects are both repairable and nonrepairable. A missile on the launching pad is repairable; in flight it is not. Yet, the value of m will differ for either stage. For the flight-stage there will be far greater environmental punishment, such as shock, vibration, and temperature, so that MTTF will be poorer than MTBF. Also, some components may not be needed during either the ground or the flight phase, thus compounding the differences. In practice, all of these factors must be considered in order to answer the question: "Will this thing work, and for how long?"

Consider an electronic guidance system. Installed in a seagoing vessel, it may have an MTBF of 1,000 hours. From Table 16-4, we find that P_s for 1,000 hours is 36.8 percent. Obviously, this is a

Table 16-4.   Reliability in Terms of Probability of Survival for Various Failure Rates and Operating Times (Based on Exponential Distribution)

Failures per 100,000 hr.*	Mean Time to Failure	Hours of Operation														
		5	10	20	50	100	200	300	500	1000	2000	3000	5000	7500	10,000	15,000
		Probability, Percent, of Survival, P,														
1	100,000	99.995	99.99	99.98	99.95	99.9	99.8	99.7	99.5	99.0	98.0	97.0	95.1	92.8	90.5	86.1
2	50,000	99.990	99.98	99.96	99.90	99.8	99.6	99.4	99.0	98.0	96.1	94.2	90.5	86.1	81.9	74.1
3	33,300	99.985	99.97	99.94	99.85	99.7	99.4	99.1	98.5	97.0	94.2	91.4	86.1	79.9	74.1	63.8
4	25,000	99.980	99.96	99.92	99.80	99.6	99.2	98.8	98.0	96.1	92.3	88.7	81.7	74.1	67.0	54.9
5	20,000	99.975	99.95	99.90	99.75	99.5	99.0	98.5	97.5	95.1	90.5	86.1	77.9	68.7	60.7	47.2
6	16,700	99.970	99.94	99.88	99.70	99.4	98.8	98.2	97.0	94.2	88.7	83.5	74.1	63.8	54.9	40.7
7	14,300	99.965	99.93	99.86	99.65	99.3	98.6	97.9	96.6	93.2	86.9	81.1	70.5	59.2	49.7	35.0
8	12,500	99.960	99.92	99.84	99.60	99.2	98.4	97.6	96.1	92.3	85.2	78.7	67.0	54.9	44.9	30.1
9	11,100	99.955	99.91	99.82	99.55	99.1	98.2	97.3	95.6	91.4	83.5	76.3	63.8	50.9	40.7	25.9
10	10,000	99.95	99.9	99.8	99.5	99.0	98.0	97.0	95.1	90.5	81.9	74.1	60.7	47.2	36.8	22.3
20	5,000	99.90	99.8	99.6	99.0	98.0	96.1	94.2	90.5	81.9	67.0	54.9	36.8	22.3	13.5	5.0
30	3,330	99.85	99.7	99.4	98.5	97.0	94.2	91.4	86.1	74.1	54.9	40.7	22.3	10.5	5.0	1.1
40	2,500	99.80	99.6	99.2	98.0	96.1	92.3	88.7	81.9	67.0	44.9	30.1	13.5	5.0	1.8	0.2
50	2,000	99.75	99.5	99.0	97.5	95.1	90.5	86.1	77.9	60.7	36.8	22.3	8.2	2.4	0.7	0.1
60	1,670	99.70	99.4	98.8	97.0	94.2	88.7	83.5	74.1	54.9	30.1	16.5	5.0	1.1	0.3	
70	1,430	99.65	99.3	98.6	96.6	93.2	86.9	81.1	70.5	49.7	24.7	12.3	3.0	0.5	0.1	
80	1,250	99.60	99.2	98.4	96.1	92.3	85.2	78.7	67.0	44.9	20.2	9.1	1.8	0.3		
90	1,110	99.55	99.1	98.2	95.6	91.4	83.5	76.3	63.8	40.7	16.5	6.7	1.1	0.1		
100	1,000	99.5	99.0	98.0	95.1	90.5	81.9	74.1	60.7	36.8	13.5	5.0	0.7			
200	500	99.0	98.0	96.1	90.5	81.9	67.0	54.9	36.8	13.5	1.8	0.3				
300	333	98.5	97.0	94.2	86.1	74.1	54.9	40.7	22.3	5.0	0.3					
400	250	98.0	96.1	92.3	81.9	67.0	44.9	30.1	13.5	1.8						
500	200	97.5	95.1	90.5	77.9	60.7	36.8	22.3	8.2	0.7						
600	167	97.0	94.2	88.7	74.1	54.9	30.2	16.6	5.0	0.3						
700	143	96.6	93.2	86.9	70.5	49.7	24.7	12.3	3.0	0.1						
800	125	96.1	92.3	85.2	67.0	44.9	20.2	9.1	1.8							
900	111	95.6	91.4	83.5	63.7	40.7	16.5	6.7	1.1							
1000	100	95.1	90.5	81.9	60.7	36.8	13.5	5.0	0.7							

*In place of hours, any time unit may be used, so long as the units are alike for the 3 terms involved: Average number of failures per period (λ); Mean time to (or between) failure(s) (m); and operating time (t). In place of a time period, a time-related basis may also be used, such as "number of actuations" or "number of flexures" until failure. For example, relay life might be expressed in terms of failure-free actuations and wire life may be measured by the number of flexures that the wire can withstand before breaking.

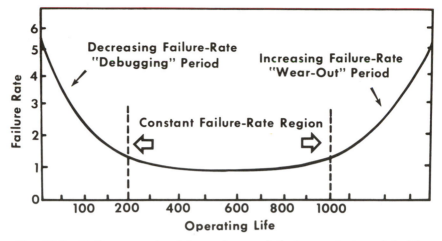

Fig. 16-5. Failure cure involving a long, relatively constant period. The exponential distribution is derived from survival probabilities associated with the constant region. Early failures (during "debugging") and the "wearout" phase are generally not considered part of the effective life of a product.

rather poor reliability. But there is a 95.1 percent probability for the system to operate properly for at least 50 hours. We may accept the guidance system under the provision that preventive maintenance occurs approximately every 50 hours of actual use. If this maintenance succeeds in removing parts—resistors, transistors, tubes— that have tended to drift toward limiting performance values, thus correcting incipient trouble spots, then there will in fact prevail a 95.1 percent reliability.

Installed in an airplane, however, the system is exposed to greater environmental stresses. Its MTBF may deteriorate by a factor of 2, so that MTBF becomes 500 hours. P_s for 1,000 is now only 13.5 percent and P_s for 50 hours is 90.5. The system may be rejected as inadequate. For the sake of illustration, however, assume that this same system is now being considered for a missile. Here the stresses are so considerable that instead of an m of 1,000 or 500, we have an m of but 100. P_s for 50 hours' survival has become 60.7 percent, and the m now refers to MTTF. But the missile may have to function for only 5 hours,[3] in which case P_s becomes 95.1. An odd situation has

[3]Since a missile can cover half the globe in only 0.5 hour, the required 5 hours may seem unrealistic. In practice, however, designers of missiles and other complex and critical devices usually allow for a safety factor. In this case a safety factor of 10 would raise the 0.5 hours to 5 hours.

developed. The component that was unsuitable for aircraft has become adequate for a missile. But usually the shoe is on the other foot: the more demanding an environment is, the more difficult the attainment of suitable products to meet requirements.

Practical Value of Reliability Information

Availability of data regarding m and P_s is of considerable benefit to management in many situations, such as the following:

Maintenance. Knowledge of the life expectancy and wear-out characteristics of component parts of equipment, with estimates of the time interval in which equipment may fail, leads to the development of (1) sound maintenance frequencies and (2) appropriate provision for spare parts, stand-by equipment, and replacements.

Maintainability plus reliability affect equipment availability. For example, if mean time between failures is 200 hours and it takes 50 hours to repair and maintain equipment for this time interval, then equipment availability is $200/(200 + 50) = 80$ percent.

Mission Success. Assessment of reliability permits evaluation of the success likelihood of a mission. Appropriate action can then be taken. For example, if the probability of success of a single space probe to a distant planet is considered inadequate, one may be able to improve the chances of overall success by dispatching two probes. For other purposes, such as a manned flight, equipment may be rejected until it has been redesigned to be suitable.

Cost Control. It is possible to correlate product reliability requirements with product costs: The quality of components used is then matched with the needs specified. Thus, a transistor going into a television set can be produced to a much lower price than a transistor going into a missile. Reliability analysis ascertains the degree of reliability needed for each purpose.

Safety. Only by knowing reliability of components can equipment be built for maximum attainable safety.

Evaluation of reliability thus becomes an important engineering and management tool, without which the age of space travel, computerization, and automation would be unthinkable; but which is equally applicable to the home appliance and automobile.

Practical Precautions

Assessment of reliability should at all times be consistent with the underlying failure distribution. Thus, while it is true that most wear-out failures tend to follow an approximate normal pattern, it is highly desirable that actual checks confirm that this distribution does apply to a particular type of product.

For chance failures, our analysis procedures have been based on an underlying exponential curve, involving constant force of failure. For a good many types of output, this procedure is valid. It is assumed, however, that the usually predominant period of constant failures has been identified, since the "debugging" stage and the "wear-out" phase are of an entirely different nature. These matters were brought out previously and are underscored further by the typical failure rate experience portrayed in Fig. 16-5. The manner in which a constant rate or force of failure gives rise to an exponential distribution of survival probabilities has been brought out.

Of course, chance failures will many times display patterns that involve forces other than those of constant failure, and thus a great variety of distribution shapes may be experienced. Fortunately, there is a general type of distribution, known as the Weibull distribution, which is capable of describing these multiform patterns. Unlike the exponential distribution, which has just one characteristic parameter, or the normal, which has two parameters (mean and standard deviation), Weibull has three additional parameters. These are known as the shape parameter, the scale parameter, and the location parameter. When the shape parameter of the Weibull curve is 1.0, it is identical to the negative exponential distribution; when the parameter is 3.5, a shape close to the normal curve results. The many forms that the distribution may assume are illustrated further in Fig. 16-6.

Finally, it should be noted that a consideration of the exponential, normal, and Weibull distributions does not exhaust the types of patterns that may, on occasion, be encountered. Our discussion, of necessity, is limited to the predominant situations likely to arise[4].

[4]Two references may interest some readers at this point. The first is James R. King, *Probability Charts for Decision Making* (New York: Industrial Press, 1971), which covers over a dozen distributions and gives practical applications. The other is a work by Buddy L. Myers and N. L. Enrick, *Statistical Functions* (Kent, Ohio: Kent State University Press, 1970), which brings mathematical-statistical derivations of these distributions in terms of elementary calculus.

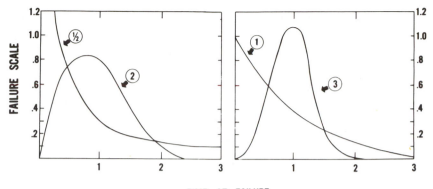

TIME AT FAILURE

Fig. 16-6. Weibull distributions. Circled values represent shape parameter. Typical values of the important shape parameter are shown. Location parameter is zero, while scale parameter is 1. When the shape parameter is 1, an exponential curve results. As the value decreases, such as to ½, the curve becomes steeper. At values above 1, a one-peaked (so-called "unimodal") distribution results. Between the shape parameter values of 3 to 3.8 a near-normal form results. Finally, as the shape parameter increases in magnitude, a more and more marked "peaking" of the curve is reflected.

REVIEW QUESTIONS

1. How is reliability defined?
2. What is the distinction between early failures, chance failures and wear-out failures?
3. What is the purpose of cumulative distributions in portraying failure frequencies?
4. If the failure rate (λ) is given as 10 failures per 10,000 hours, what is the mean time to failure (MTTF or m)?
5. Given an m of 4,000 hours, what would be the number of failures per hour?
6. In general terms, what is the effect of constant force of failure on the resultant failure distribution (reliability curve)?
7. Given 4 failures per 100,000 hours of operation, find (A) MTBF and (B) P_s for 100, 200, 500 and 1,000 hours.
8. A system with MTBF of 200 hours is used for only 100 hours before it undergoes a thorough overhaul. What is the system's effective P_s?
9. A sample of neon bulbs for a computer console is analyzed on the basis of a repair-reporting system. The following are the hours to failure: 4,940, 5,110, 6,050, 5,060, 4,400, 5,600, 4,950, 4,100, 4,800, 4,990. What is (A) MTTF, (B) failure rate per 1,000 hours, (C) P_s for 5,000, 7,500 and 10,000 hours?
10. What is the maintainability of a system which requires 100 hours of repair and maintenance downtime for every 400 hours of operation?

11. Assume that a system has a maintainability of 0.8 or 80 percent and that P_s is 0.7. You are given a formula:

Effectiveness = Reliability × Availability

What is the effectiveness in percent?

12. An orbiting system consists of the three principal components of receiver R, analyzer A, and transmitter T, with 10, 50, and 100 failures per 100,000 hours. The receiver must function properly for 1,000 hours while analyzer and transmitter will be needed only one-fifth of that time. In order to determine P_s of the system, we must first express the data in terms of MTTF and then determine P_s for each component. In this determination, reliability values should be based on the anticipated time of operation of each of the three units per 1,000 hours. What is the system reliability so ascertained?

For additional review materials, see Cases 13 and 14 in Appendix 3.

CHAPTER 17

Reliability Assurance

Having discussed the fundamental concepts in assessing reliability, we must now turn our attention to a further vital aspect: How do we assure that products have the required degree of reliability for their intended mission? Engineering proof of such reliability must be marshalled in order to support reliability statements.

Reliability Testing

Assurance of reliability begins with reliability testing. One type of such testing has already been discussed. Sample pieces of product, selected randomly from a production lot or other rational group, are subjected to life testing. The number of failures are recorded along with the age at failure. The electron tube and the transistor examples are among the products subjected to such tests. In many practical situations, however, it is not feasible from a standpoint of time, manpower, testing facilities, and costs to make extensive observations.

Abbreviated Life Tests. On many types of products, the characteristic force-of-failure pattern can be predicted with reasonable validity. When we are dealing with chance failures and thus a constant force of failure, the data of Table 16-4 will properly predict reliability on the basis of a brief initial test.

Failure-Repair Runs. Although an item may be nonrepairable in actual application, it is often feasible to repair failures during reliability testing. Fewer test units will thus be needed since a repaired item can be looked at as a "new" unit; or else the value of MTBF of testing can be considered an MTTF for end-use purposes.

Accelerated Tests. Ingenious ways of accelerated testing to compress time have been found. One way is to change time, such as hours, to some other criterion, such as number of actuations. For example, instead of testing a snap-action thermostat for two years, we can rig up a device which through rapid temperature changes causes many tens of thousands of "on" and "off" actuations in a few days. The failures observed are now MTTF in terms of number of actuations rather than hours, but Table 16-4 and Fig. 16-4 will be equally valid. We can also convert the actuations observed to hours. For example, if in its end use a thermostat has an average of 10 actuations per hour, then a MTTF of 100,000 actuations corresponds to 10,000 hours of operation.

Test of Increased Severity. Sometimes it is possible to increase the severity of the required temperatures, vibrations, current strengths and other stresses on a test specimen or device. If the item does not fail during relatively brief exposure to these intensified factors, extrapolations can be made to predict the life of the object under lesser but prolonged periods of stress. Because of the complexities and pitfalls in such extrapolations, the approach must be employed with great care and circumspection.

Tests of Large Sample Sizes. Instead of testing a few items to destruction, we can test many items until a few have failed. For example, instead of testing 10 capacitors until they fail, we can test 100 of them until 10 have failed. The time savings accomplished are significant, but the costs of sampling large numbers of units are often considerable.

The last mentioned approach requires the application of a sampling table, as will be discussed next.

Application of Sampling Table

The use of the sampling plans in Table 17-1 requires an illustrative example. Assume that a shipment of transistors has been received, and that a mean life of 10,000 hours has been specified. Only about 200 hours are available for testing. The ratio of this testing time, 200, to the specified mean life, 10,000, is 0.02. Entering the Table for this column, we find that if a sample of 116 resistors is tested for 200 hours and zero failures occur, we can be 90 percent confident that the shipment as a whole conforms to the specified mean life of 10,000 hours.

Statistically, this is a sound approach. But if one failure is found, calling for a rejection of the lot, we might receive squawks from the supplier. "After all, one failure is possible even under good production," he may claim. So, we may decide that we need to find at least two failures for rejection, meaning that for acceptance up to one

Table 17-1.   Sample Sizes Needed for Reliability Testing*

Ac-cept-ance No.	Ratio of Testing to Specified Mean Life†												
	1.0	0.5	0.2	0.1	0.05	0.02	0.01	0.005	0.002	0.001	0.0005	0.0002	0.0001
	Sample Size, No. of Units												
0	3	5	12	24	47	116	231	461	1,152	2,303	4,606	11,513	23,026
1	5	9	20	40	79	195	390	778	1,946	3,891	7,780	19,450	38,898
2	7	12	28	55	109	266	533	1,065	2,662	5,323	10,645	26,612	53,223
3	9	15	35	69	137	333	668	1,337	3,341	6,681	13,362	33,404	66,808
4	11	19	42	83	164	398	798	1,599	3,997	7,994	15,988	39,968	79,936
5	13	22	49	97	190	462	927	1,855	4,638	9,275	18,549	46,374	92,747
6	15	25	56	110	217	528	1,054	2,107	5,267	10,533	21,064	52,661	105,322
7	16	28	63	123	243	589	1,178	2,355	5,886	11,771	23,542	58,855	117,710
8	18	31	70	136	269	648	1,300	2,599	6,498	12,995	25,990	64,974	129,948
9	20	34	76	149	294	709	1,421	2,842	7,103	14,206	28,412	71,030	142,060
10	22	37	83	161	319	770	1,541	3,082	7,704	15,407	30,814	77,034	154,068
11	23	40	89	174	344	830	1,660	3,320	8,300	16,598	33,197	82,991	165,982
12	25	42	95	187	369	888	1,779	3,557	8,891	17,782	35,564	88,908	177,816
13	27	45	102	199	393	947	1,896	3,792	9,479	18,958	37,916	94,790	189,580
14	29	48	108	212	417	1,007	2,013	4,026	10,064	20,128	40,256	100,640	201,280

*Entries show the minimum sample size that must be tested for the length of time shown, to assure a mean life (MTTF or MTBF) of at least the length required, with a confidence of 90 percent that this mean life actually prevails in the lot accepted.

†Table is based on exponential failure distribution.

Source: M. Sobel and J. A. Tischendorf, *Proceedings of Fifth National Symposium, Reliability Quality Control,* IRE (January, 1959), pp. 108-118.

failure is permissible. However, as the table shows, our sample size now increases to 195, and the practical inspector, engineer, and manager may be dismayed at this relatively large sample size. Lot-by-lot acceptance of products that must meet stringent requirements is indeed a costly business in terms of materials, manpower, and testing time.

Sequential Sampling Plans

It is also possible to employ the sequential sampling plans of Table 2-1, pages 14 and 15. Assume, for example, that a shipment of 2,000 flashlight batteries has been received. The specification states that any battery must have a life of at least 3 hours, and that the allowable percent defective shall be no more than 5.0. The table shows that our first sequential sample must consist of 50 units. If none fail, the lot is accepted. If 6 fail the lot is rejected. For 1 to 5 failures, further sampling is called for. The procedure is described in detail in Chapter 2, and the sampling risks involved are discussed in Chapter 3.

Verification of Distributions

What is the underlying failure distribution for a particular type of

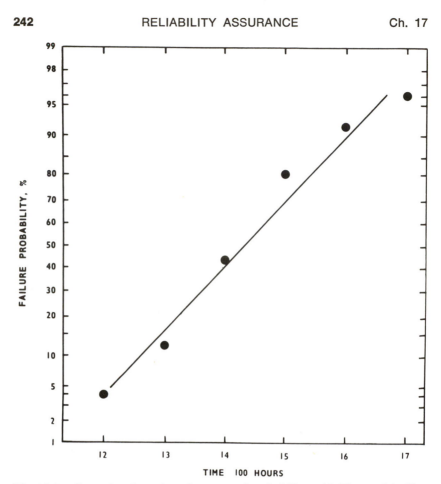

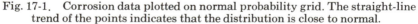

Fig. 17-1. Corrosion data plotted on normal probability grid. The straight-line
trend of the points indicates that the distribution is close to normal.

product undergoing a given kind of life test? Although technical and
other considerations may permit valid assumptions, it is generally
best to verify the distribution by actual checks. Usually, this task is
accomplished by cumulating the test frequencies and making plots
on applicable probability paper.

The corrosion failures observed on a sample of 24 components, as
listed below, are graphed on the probability grid of Fig. 17-1.

The reliability engineer had assumed that a normal curve would
be applicable, and the plot accordingly is on normal probability
paper. The straight free-hand line drawn to the points fits relatively

Table 17-2. Corrosion Failures Observed for a Sample of 24 Units

(a)	(b)	(c)	(d)	(e)
		Frequency		Plotting
Time to Failure, Hours	Observed No.	Cumulative		Frequency*
		No.	%	%
1,200	1	1	4.2	4
1,300	2	3	12.5	12
1,400	8	11	45.8	44
1,500	9	20	83.3	80
1,600	3	23	95.8	92
1,700	1	24	100.0	96

*Plotting Frequency $= 100 \times (c)/(n + 1)$
where (c) is given in column (c) above and n is the sample size. The use of $(n + 1)$ for the plotting frequency represents, in effect, an adjustment to the cumulative percentage in column (d).

well. When such a linear relationship occurs, this serves as confirmation that the assumed distribution does prevail.

The purpose of the adjustment of the cumulative frequencies to obtain the plotting points in percent will now become apparent. Without this so-called "unbiasing correction," any final point would *always* be at 100 percent, thereby either producing an undue effect on the line of average relationship, or else causing the final point always to be considerably distant.

Value of Validation

Benefits accruing from graphic verification of the pattern of failure rates are numerous. Reliability predictions and extrapolations will be based on an understanding of the failure distribution. Sampling plans can be chosen to be compatible with the underlying probabilities. Finally, such statistical measures as the mean and variance can be evaluated quickly.

For a normal curve, the arithmetic mean on probability plots corresponds to the failure scale value at the point where the fitted line crosses the 50-percent level—which is 1,430 hours for our illustration. The value is approximate. An exact calculation would yield 1,460 hours for the mean. The slight discrepancy is ascribable to (1) variations in judgment when drawing the free-hand line of fit and (2) lack of full normality of the data (otherwise the points would fall on and not around the line).

The standard deviation is obtained similarly. Recall that the

Table 17-3.   Mileage to First Major Repair of Service Vehicles

(a)	(b)	(c)	(d)
Time to First Major Repair, in 1,000 Miles	Number of Vehicles	Cumulative Number	Plotting Percentage $= (c)/(n+1)$ *
4.25	1	1	10
4.95	1	2	20
5.75	1	3	30
6.85	1	4	40
7.85	1	5	50
9.35	1	6	60
10.95	1	7	70
13.10	1	8	80
16.85	1	9	90

The columns (b), (c), (d) fall under the spanning header **Frequency**.

*Sample size $n = 9$, so that $(n + 1) = 10$. We use our previously established formula, Plotting Percentage = (Cumulative Frequency)/$(n + 1)$

population mean ± 1 standard deviation, representing a range of 2 standard deviations, includes from 16 to 84 percent of the distribution values. Accordingly, we locate the 16- and 84-percent levels on the probability scale, move horizontally until the line of fit is reached, and note the corresponding base scale values. These are 1,300 and 1,570 hours respectively. The difference of 270 hours represents 2 standard deviations. Hence, $270/2 = 135$ hours is the value for the standard deviation, which is almost identical to the calculated value.

Although our discussion has been in terms of the normal curve, it should be noted that probability paper is available for all types of distributions, permitting the evaluation of sampling data in a manner parallel to the methods given here.[1]

Coping with Small Sample Sizes

Ordinarily, when the sample size is small, an estimate of the underlying distribution cannot be obtained. This is not so with probability papers. Even when a mere seven (or, in extreme instances, five) data points are available, a plot can be made and a valid estimate of the applicable parameters be made. Illustration of

---

[1]This topic deserves a book in itself. The reader may wish to refer to James R. King, *Probability Charts for Decision Making* (New York: Industrial Press, 1971).

this principle is provided in Table 17-3 by the example of how many miles nine service vehicles ran before each required major repair.

Since the failure pattern is unknown, we shall plot the failures on Weibull probability paper, which includes the exponential and near-normal cases (depending on shape parameter).

Although a first graphing of the points results in a curve, in Fig. 17-2, we should by trial discover a constant which is subtracted from each failure observation, seeking to obtain a straight trend. In our example, the constant turns out to be 3, which is thereby the estimated location parameter. Having established linearity, we also confirm the validity of using Weibull grid. Next[2]

1. A perpendicular, from the fitted line to the estimation point (arrow), crosses the shape parameter scale at 1.3.
2. The next higher scale, crossed at close to 60, represents the cumulative failures up to the mean. A horizontal line at that level of the ordinate crosses the fitted trend at 5.5 for the base scale. Add the location parameter 3 to obtain 8.5 for the average mileage in 1,000s.
3. The dotted horizontal (preprinted) line crosses the fitted trend at 6.5, which is again increased by 3 to obtain the scale parameter 9.5.

Salient results are that (1) a Weibull distribution prevails, (2) 60 percent of the vehicles will have failed when average life expectancy of 8,500 miles is reached, and (3) the important shape parameter is 1.3. Now from Fig. 16-6 we note that at 1.0 we would have an exponential curve. Since 1.3 is higher, we have a unimodal (one-peaked) curve. Moreover, the relatively small value of the shape parameter indicates that the curve is peaked to the right and then tails out toward the right.

Assurance Value of Probability Plots

The practical uses of graphed distributions are manifold. Among them are:

1. Verification of the underlying frequency pattern guides the selection of appropriate, consistent, acceptance sampling plans.

---

[2]The particular type of approach given is due to Dr. Lloyd S. Nelson, Manager, Applied Mathematics Laboratory, General Electric Company. His cooperation in developing the present tutorial example is gratefully acknowledged.

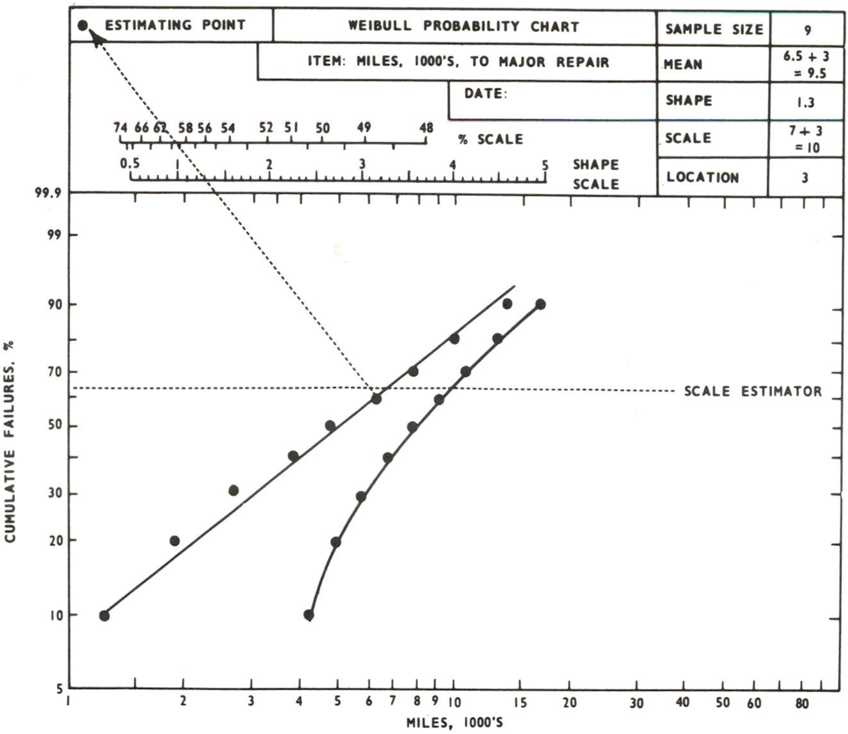

Fig. 17-2. Weibull plot. Curve-forming points become straight-trended by subtraction of the location parameter 3, found by trial. Dotted diagonal line is drawn in to find shape and percentage scale values. Horizontal dotted line is preprinted.

2. Proportions of units failing at a particular age are estimated by means of interpolation along the trend line.

3. Extrapolation is facilitated. After the seventh failure, for example, at which juncture 70 percent of the vehicles had been involved and the corresponding mileage was 11,000, extension of the trend line would have brought us immediately to the estimate of approximately 87 percent failures at 15,000 miles (remember to add the location parameter of 3,000 miles to the base scale value for the linear trend).

4. A small sample will nevertheless give an adequate estimate of the entire distribution of failures.

5. Closeness of fit of observed to theoretical distributions is possible, by comparing the trend line against the plotted points.

6. Projections are possible—for example, the anticipated failure rates if certain types of failure causes are modified or removed.
7. Cost effects of warranties, in terms of percent of units requiring field service or likely to be returned, can be evaluated.

In the present instance, since the shape parameter is close to 1.0, we may feel justified in basing acceptance sampling and other estimates on the negative exponential distribution. If greater accuracy is desired, however, special Weibull tables (more complex to use) may be applied.

Methods for Incomplete Tests

Reasons of practicality and economy often dictate a termination of testing and observations before all of the units have failed. A manufacturer had designed a special drive motor, which had been placed into production gradually, as the need arose. Over a period of time, ten units had been placed. Five had burned out within periods ranging from 1,500 to 10,000 hours, while another five units were still operational after from 3,050 to 12,000 hours, as noted in columns (a) and (b) below. The steps leading to the values in (c), (d), and (e) are self-explanatory.

Fitted on Weibull chart, the data form a straight plot (see Fig.

Table 17-4. Operating Times and Failure Incidents of Drive Motor

(a)	(b)	(c)	(d)	(e)
Hours of Operation*		Time Ranking	Failure Hazard, Percent	
At Time of Failure	At Termination of Test	(Longest Time of Operation Receives Rank 1)	$(= 100/c)$†	Cumulative Hazard Value Found‡
1,500	. . .	10	10.0	10.0
. . .	3,050	9	. . .	. . .
3,500	. . .	8	12.5	22.5
4,100	. . .	7	14.3	36.8
. . .	5,750	6	. . .	. . .
. . .	6,300	5	. . .	. . .
7,000	. . .	4	25.0	61.8
. . .	8,450	3	. . .	. . .
10,000	. . .	2	50.0	111.8
. . .	12,000	1	. . .	. . .

*At the end of the test observations, a motor had either failed—at some prior, recorded time shown in column (a)—or else it was still running—having clocked the running hours in column (b).

†For example, in the first row $100/c = 100/10 = 10$. In the third row, $100/8 = 12.5$.

‡Column (e) is obtained simply by cumulating the hazards of the prior column. The first row remains 10, the third row becomes $10 + 12.5 = 22.5$, the fourth row is $22.5 + 14.3 = 36.8$.

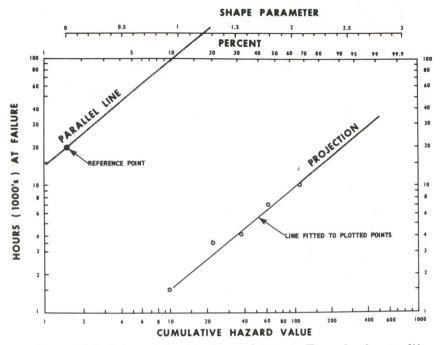

Fig. 17-3. Weibull hazard chart for incomplete tests. Example of motor life involving 10 units, only five of which had burned out when observations were terminated.

SOURCE: This type of Weibull grid was developed by Dr. Wayne Nelson. *Quality Technology*, vol. 1, no. 1. pp. 27–52 January, 1969. and is commercially available only from TEAM, Box 25, Tamworth, N.H. 03886 under General Electric Company license.

17-3). A parallel dotted line, through the preprinted reference point, reveals a shape parameter of approximately 1.2. Reliability assurance statements are now obtainable with the aid of the percentage scale (top). For example, if future units of the same design and manufacture are guaranteed for 4,000 hours, how many free renewals may have to be provided? Entering the vertical scale at 4,000 hours, we cross the diagonal fitted plot at a level corresponding to approximately 30 percent on the upper scale. Out of every ten motors sold, then, three will require replacement.

Nominal life can also be ascertained. Enter the top scale at 50 percent, moving down until the diagonal is reached at a level corresponding to 7,000 hours. It is thus apparent that one does not have to wait until all units under test have failed to obtain reliability information.

Practical Reliability Assurance

This chapter has been concerned with the principal aspects of reliability assurance—i.e., with the testing of products for reliability in accordance with specified requirements. In presenting methods of sampling and testing, we have pointed out necessary precautions in interpreting test results. A comprehensive assurative testing program must rely on many tools:

1. Complete descriptions are needed of the products that undergo reliability testing. These descriptions will include specifications for both quality and reliability.
2. Concise prescriptions should be made for the performance of tests, including meticulous attention to the ambient conditions—such as number of operating cycles and times, temperatures, shock, pressures and vibrations—that are to prevail during testing.
3. Sampling procedures, sample sizes, criteria for judging the success or failure of an item, and acceptance and rejection values for action on a lot should be definitely laid down.
4. Knowledge of the calculated sampling risks, such as embodied in operating characteristic curves or tabulated data on the probabilities of sampling errors.

It will often be impractical to subject a large and complex system to performance testing. There will, however, be data on component reliabilities. Fortunately, it is possible to synthesize from these individual probabilities the expected reliability of the system as a whole, as will be shown in the next chapter.

In reliability assurance, just as in quality control, we thus observe that a sound program provides for strategic testing at various stages of production. When such testing occurs, we may at times be at a loss in properly defining it. For example, is a check of tensile strength of a critical structural part a quality control or a reliability assurance test? From a practical viewpoint, we resolve this problem by observing that reliability is but one—albeit a most important—characteristic of good product quality. Indeed, installation of reliability assurance often has been most successful in those organizations that have already had a good quality control program, with detailed quality assurance procedures,[3] in effect.

---

[3]For illustrative examples see C. P. Covino and A. W. Meghri, *Quality Assurance Manual* (New York: The Industrial Press, 1967).

Despite good testing, conscientious control, and effective overall programs, the rate of actual failures in the field will in many instances exceed expectations. The reasons for such problems are numerous. For example, conditions encountered in use of equipment may be more severe than anticipated, laboratory testing cannot fully duplicate environmental stresses, people may fail to handle equipment as specified for proper performance, and maintenance work may be inadequate. While these types of trouble cannot be avoided, there will be reporting systems that lead to the build-up of field-failure data. The analysis of the various types of failure, in turn, leads to the identification of weak spots in the product cycle— from design through production and service usage. Review and revision of designs, adjustments in processing, and improvements in service instructions will result from these analyses. As a consequence, a continuing reliability growth of equipment occurs.

Effective reliability assurance is thus a never-ending program.

REVIEW QUESTIONS

1. Among the various types of reliability testing available, we may note abbreviated life tests and failure-repair runs. What are the principal features of such tests?
2. Other types of reliability testing are accelerated tests, tests of increased severity, and tests of large sample sizes. Under what conditions would you use each of these?
3. A product has an exponential failure distribution. The planned ratio of testing time to specified mean life is 0.05 and the sample size is 190 units. What is the appropriate acceptance number?
4. A minimum MTTF of 1,000 hours has been specified. Failures are expected to show an exponential distribution. We have time to test for 100 hours. The acceptance number is 1. We desire a confidence level of 90 percent that the mean time to failure of 1,000 is the actual life prevailing in the lot. What is the sample size that should be tested?
5. For acceptance sampling, assuming an exponential failure distribution, what are the specific factors that must be considered in arriving at proper reliability testing?
6. In what ways is reliability testing similar to lot-by-lot sampling inspection?
7. Aside from reliability testing, what other factors are important in setting up a sound reliability assurance program for a particular product?
8. For the analysis of wearout failures data tabulated in the preceding

chapter, what are the plotting frequencies (for graphing in probability paper)?

 9. What are the plotting frequencies for the repair-reporting data of
    review question 9 in Chapter 16?
10. Note the corrosion failure data (which should be plotted on Weibull
    paper) for a sample of 24 units, on page 243. What are the mean,
    shape, scale, and location parameters? Why can one always plot
    normal distribution data on Weibull paper?
11. When the shape factor of a Weibull distribution is 1.0 or smaller,
    the failure rate decreases with time. How would you verify this
    statement?

See Case 15, Appendix 3, for additional review materials. If frequency
distributions were prepared for Cases 2, 3, and 8, these can now be further evaluated for normality by means of probability paper.

CHAPTER 18

Reliability Design

Important as such assurative activities as reliability testing and evaluation are, they cannot enhance the ultimate soundness of the product itself. Accordingly, a great deal of emphasis in reliability engineering has been placed on the reliability inherent in the design itself. In turn, the statistical methods of reliability analysis aid the design-engineer in his work, by predicting the reliability to be expected from various alternative designs. On the basis of these predictions, an optimal decision can then be made.

Precision of predictions is essential; otherwise inadequate or overdesigned product may result. An inadequate door hinge will cause complaints and returns from premature failures, while an overdesigned hinge may incur undue costs of heavy-gauge steel, corrosion-resistant alloys, and excessive machining, thus pricing the product outside the market. In space craft, problems of precise reliability prediction become critical. Underdesigned components may be catastrophic, while overdesign may bring burdensome problems of weight and complexity.

Reliability Improvement Through Redundancy

The nonredundant system in Fig. 18-1 consists of three elements; a transmitter, a receiver, and a coder. The probabilities of failure-free operation for 100 hours of this system depend on the component reliabilities.

Using the symbol P_s to indicate "probability of successful operation" for the interval of 100 hours, we observe these component or element reliabilities:

$$P_s \text{ (Transmitter)} = P_s \text{ (T)} = 70 \text{ Percent or } 0.70$$
$$P_s \text{ (Receiver)} = P_s \text{ (R)} = 90 \text{ Percent or } 0.90$$
$$P_s \text{ (Coder)} = P_s \text{ (C)} = 80 \text{ Percent or } 0.80$$

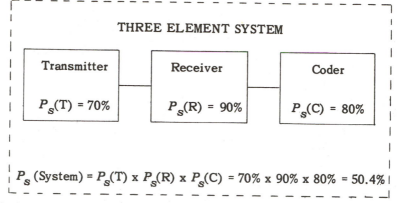

Fig. 18-1.   Illustration of nonredundant system. There are three elements in the system: a transmitter, a receiver, and a coder. Each performs its specific function, and there is no duplication of these functions.

We read P_s (transmitter) as "probability of success of transmitter" in sending out the message. Similar reading applies to all other such uses of the parentheses after a probability statement.

It is apparent that the reliability of the system hinges on the proper operation of all three elements. Failure of one defeats the entire system. Therefore,

$$P_s \text{ (System)} = P_s \text{ (T)} \times P_s \text{ (R)} \times P_s \text{ (C)}$$
$$= 0.70 \times 0.90 \times 0.80 = 0.504$$
$$= 50 \text{ percent, approximately}$$

The reader will rightly observe that this is a relatively poor reliability. Indeed, for purposes of simplified arithmetic the example has involved rounded, rather low reliabilities. A more realistic set of reliabilities would have been 0.97, 0.99, and 0.98 in place of the three probabilities just given. However, even with those higher values, the P_s of the system would be only $0.97 \times 0.99 \times 0.98$ or 94 percent, which may be considered inadequate for many purposes. Assuming that the reliability of each element cannot be improved, then by utilizing one or more extra or duplicating components—that is, the inclusion of redundancy—we might enhance overall system reliability. Let us see how this works.

Duplication of one element, such as the transmitter, would represent an instance of redundancy, as shown in Fig. 18-2. The two

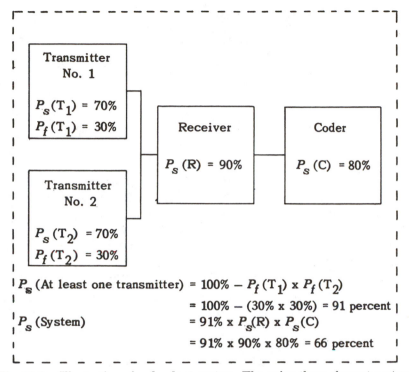

Fig. 18-2. Illustration of redundant system. The prior three-element system has been enlarged to include a duplication of the transmitting function. Two transmitters are now in the system, operating in parallel: If one transmitter fails, the other may still work. The receiver and coder are in series. If either fails or if both transmitters fail, the system is inoperative.

transmitters are in parallel, so that only in the case of failure of both will a system failure occur. What is the reliability of the transmitting subsystem? We proceed as follows:

1. Since the subsystem can only be in one of two states, "successful operation" or "failure," the two probabilities of P_s and P_f must total to 1.0 or 100 percent.
2. The probability of a failure, P_f, of either transmitter is 0.3 or 30 percent. Therefore, the chance that both will fail is 0.3×0.3 or $0.09 = 9$ percent.
3. Subtraction of this 0.09 from 1.00 gives P_s of 0.91 or 91 percent for the transmitting subsystem.

The subsystem, in turn, is in series with the receiver and the coder. For the system as a whole, therefore, we have:

$$P_s \text{ (System)} = P_s \text{ (Subsystem)} \times P_s \text{ (R)} \times P_s \text{ (C)}$$
$$= 0.91 \times 0.90 \times 0.80 = 0.655$$
$$= 66 \text{ percent}$$

Redundancy has increased overall system reliability from the original 50 to a 66 percent reliability. Further improvements may be possible, through receiver and coder duplication, or through employment of three rather than just two elements in parallel. The price for increased reliability through redundancy will be in terms of the extra cost of the duplication of facilities. Complexity of the system as a whole may increase even further, such as when "fail-safe" devices are required that automatically switch operations from failing to working elements in a subsystem.

Reliability in Single- and Double-Limit Operations

In the prediction of reliability of the transmitter-receiver-coder system, we were concerned with only one possible type of failure: an element might not function when required. This is a single-limit type of failure. A double-limit operation involves two types of malfunction. Take, for example, a switch. There is always a possibility that because of "stickiness" or a weak spring or other reasons, the switch might fail to close when required. There is, however, also a second type of malfunction. The switch might close prematurely. A slightly loose assembly, affected by further stresses of low temperatures, jarring, and rapid acceleration, might cause a switch to shut when no closure of a circuit is intended. Thus, there are two types of failure: nonclosure and premature closure.

If, instead of a switch we had been dealing with a valve, we might again be facing two types of possible failures: Nonclosure of a flow of liquid in a fuel system or premature shutting off of the flow. A variety of instances of two-sided failures can occur in most complex systems. A three-sided failure mode involves, for example, an explosive fuze that (1) might not light, (2) might light prematurely, (3) might explode on lighting.

In the design of equipment, it is essential that proper consideration be given to the types of failure that can occur, since each type might have a different significance. For example, a switching fault that causes a premature firing of a missile may be worse than a failure to fire.

Redundancy and Reliability of Double-Limit Operations

A large-scale system contains an element which is placed into action by the closure of a circuit between the points "a" and "b" of the diagram in Fig. 18-3. Experience has shown that a certain type of switch has a reliability of 70 percent of at least 1,000 proper closures. Considering this probability inadequate, we have placed two identical types of switches, A and B, in parallel. As a result (see calculation in Fig. 18-3), the reliability of the switching system has been increased to 77 percent.

At the same time, the likelihood that a switch might fail to close, P_f, which is 20 percent for a single unit, has decreased to $(0.20)^2 =$ 4 percent for the combination. Unfortunately, however, premature closure has increased from $P_p = 10$ percent for one switch to $P_p = 19$ percent for the two switches in parallel. If premature closure is a potentially serious hazard in terms of safety and costs, then redundancy has certainly not improved matters where it counts most heavily.

Another design, in Fig. 18-4, employs four switches, which reduces the premature failure possibilities. But we may now be dissatisfied with the increase in the probability of nonclosure. Possible designs thus far considered involve these probablities:

	Reliability and Failure Probability Percentages		
	One Switch Only	Two Switches in Parallel	Four-Switch Arrangement
Successful Closure	70	77	88.6
Failure to Close	20	4	7.8
Premature Closure	10	19	3.6

Redundancy, obviously, is not an unmixed blessing. It is precisely this fact, however, which makes analysis and prediction of reliability so important. With this statistical technique, we can inform ourselves of the relative merits of several possible designs. One can then choose the optimum of a set of alternative designs. As a still further possibility, we might develop additional and probably more complex designs. Whether or not such action is needed depends on

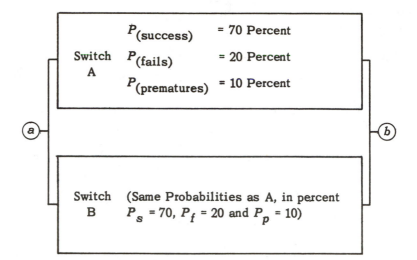

CALCULATION OF RELIABILITIES

Possible Outcomes		Probabilities of Outcomes, Percent				System
Switch A	Switch B	Switch A	Switch B	Calculations	Result	Probability
Success	Success	P_s	P_s	70 x 70	49	Success
Success	Failure	P_s	P_f	70 x 20	28	Success
Failure	Success	P_f	P_s	20 x 70		
Failure	Failure	P_f	P_f	20 x 20	4*	Non-closure
Success	Premature	P_s	P_p	70 x 10	14**	
Premature	Success	P_p	P_s	10 x 70		
Failure	Premature	P_f	P_p	20 x 10	4**	Premature Closure
Premature	Failure	P_p	P_f	10 x 20		
Premature	Premature	P_p	P_p	10 x 10	1**	
All (total)					100	

*Probability of Failure to Close is 4 Percent for the System.
**Probability of Premature Closure is 14 + 4 + 1 = 19 Percent for the System.
Therefore, System Reliability is 100 − 4 − 19 = 77 Percent.

Fig. 18-3. Reliability of a set of parallel switches. While the probability of nonclosure, P_f, is less than for a single switch (4 vs. 20 percent), the likelihood of premature closure, P_p, has increased from 10 to 19 percent).

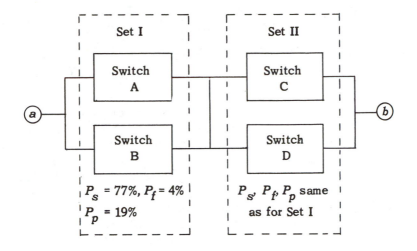

Calculation of Reliabilities

Possible Outcomes		Probabilities of Outcomes, Percent				System
Set I	Set II	Set I	Set II	Calculations	Result	Probability
Success	Success	P_s	P_s	77 x 77	59.29	
Success	Premature	P_s	P_p	77 x 19	14.63	88.55
Premature	Success	P_p	P_s	19 x 77	14.63	
Failure	Failure	P_f	P_f	4 x 4	0.16	
Failure	Premature	P_f	P_p	4 x 19	0.76	
Premature	Failure	P_p	P_f	19 x 4	0.76	7.84
Failure	Success	P_f	P_s	4 x 77	3.08	
Success	Failure	P_s	P_f	77 x 4	3.08	
Premature	Premature	P_p	P_p	19 x 19	3.61	3.61
All (total)						100.00

NOTES: Sets I and II may be considered subsystems of the new system of two pairs of switches.

The probability of successful closure of the system, P_s, is 88.55 or simply 88.6 percent as shown above.

The probability of premature closure of the system, P_p, is 3.61 and the probability of nonclosure of system, P_f, is 7.8 percent, as obtained above.

Fig. 18-4.   Reliability of two pairs of parallel switches, each pair arranged in series. Each set of switches has the reliabilities previously calculated for the single set of switches in parallel.

the exigencies of the missions envisioned for the equipment and the resources available in terms of time, money, and manpower.

Practical Reliability-Redundancy Studies

In practical reliability prediction of various possible product designs, we must recognize a number of precautionary aspects:

1. Parallel redundancy cannot be taken for granted. Equipment is often arranged in parallel, but operates as though in series. Capacitors are a prime example. They may be wired parallel but act in series in the sense that failure of either one of the two capacitors will cause malfunction of the system depending on them.

2. Combination of reliabilities depends on proper time bases. In the transmitter-receiver-coder example, as a case in point, each P_s value was based on like time values of 100 hours of operation. In many systems, however, it may be found that various elements work for only a part of the entire system's operating life. Proper cognizance of time bases must then be taken.

3. The sum of the calculated probabilities must come out to 100 percent. In the just-noted one, two, and four-switch arrangement, there were three possible outcomes per arrangement (successful closure, failure to close, and premature closure), and for each column the probabilities totaled 100 percent. If such a result is not attained, then we have (a) omitted a possible outcome, (b) misinterpreted the design or (c) made a calculation error (excepting, of course, that small deviations as a result of rounding effects are permissible).

In summary, a combination of engineering knowledge of components plus a statistical assembly of probabilities are the ingredients that jointly guide us to valid reliability predictions. Once these practical precautions have been observed, wide applicability will be found for the methods shown here, regardless of whether a large-scale system, a subsystem, or merely an assembly is analyzed. Any configuration of basic elements is predictable, once the individual reliabilities are known.

Conversion of Time Bases

It has been emphasized that time bases for reliability figures must be comparable. For example, assume that a radar defense installation includes two systems: a search unit and a tracking unit. The first must operate continuously, the second only 10 percent of the time. Assume now that, based on 1,000 hours, we have these reliabilities:

Search radar unit, $P_s = 90.5$ percent
Tracking radar unit, $P_s = 40.7$ percent

The tracking radar, however, is needed only for 100 hours out of every 1,000 hours of search. For 900 hours it is silent. Converting P_s (1,000 hours) $= 40.7$ percent to the more applicable P_s (100 hours), we find a probability of 91.4 percent. The value is obtained from Table 16-4 in Chapter 16, using the 1,000 and 100 operating hour columns at the level of 90 failures per 100,000 hours.

What is the probability that both search and tracking radar units will function when needed?

Search　　$P_s = 0.905$
Tracking　$P_s = 0.914$
P_s (System) $= 0.905 \times 0.914 = 0.827$

We thus expect an 82.7 percent probability that, when both radar units are needed, they will both be operating properly. In practice, the defense of the installation may also require proper functioning of a computer, launchers, guidance systems, and missiles. The addition of these elements in the reliability calculation will again require comparable time values and probabilities.

Component Reliability Data

The reader may ask: "How do we know the reliability of the basic components involved?" In practice, a large scale system is built up from many items, such as steel, plastic, and electronic parts. Effective prediction of system reliability depends on valid data on components' reliabilities. Many sources of information are available, compiled by trade, scientific, and technical groups and by the government.

Information has been obtained both by laboratory life tests and by gathering data on failure of equipment and parts in the field. Furthermore, within a company, information accumulates continually pertaining to the reliability of the components or other building blocks that are critical for proper functioning of equipments and other systems.

Admittedly, there is a flaw in the use of published reliability data. For example, failure rates for a certain type of transformer have been given as 20 per 100,000 hours, corresponding to a mean time between failures of 5,000 hours. But this information may not be applicable to the transformers supplied by a given subcontractor. Moreover, the failure rate may also depend on the quality of preventive maintenance in the field. As a result, the "average quality of maintenance" and the "average quality of transformer" assumed in the published failure rates may be inadequate. In time, however, as unreliable suppliers become known and are avoided and as inadequate maintenance practices are uncovered and remedied, a gradual improvement or so-called "reliability growth" will occur. Included in this growth will be the discovery within a manufacturing plant of components, design factors, use-factors, and other critical items in need of review, revision, and improvement.

Dynamic Programming

Dynamic programming principles, such as previously applied to the problem of most efficient tolerance combinations, also serve to optimize systems reliability. Refer to the transmitter-receiver-coder set with component reliabilities, P_s, of 0.7, 0.9, and 0.8 respectively and corresponding failure probabilities P_f equaling 0.3, 0.1, and 0.2. Overall reliability is to be improved by means of redundancy, but the total system is to contain no more than six components, because of space and cost limitations.

For the coder, for example, successive values for redundance are as follows:

Number of Units in Parallel	Unit P_s	Unit P_f	Calculations	Combined P_s
1	0.8	0.2	$1 - 0.2^1$	0.8
2	0.8	0.2	$1 - 0.2^2$	0.96
3	0.8	0.2	$1 - 0.2^3$	0.992
4	0.8	0.2	$1 - 0.2^4$	0.9984

For the receiver, the corresponding P_s values are 0.9, 0.99, 0.999, and 0.9999, while for the transmitter we have 0.7, 0.91, 0.973, and 0.9919 respectively, representing reliability with 1, 2, 3, and 4 units in parallel.

A system with two coders and one transmitter and receiver each, would thus show a reliability P_s of 0.96 × 0.7 × 0.9 or 0.61.

Our redundancy reliabilities are now fed into a two-phase dynamic programming scheme, as shown below.

DYNAMIC PROGRAMMING SEARCH FOR OPTIMUM RELIABLIITY

Phase 1: Analysis of Coder and Receiver Operation

Number of Coders in Parallel	Coding Operation Reliability	Number of Receivers Used in Parallel and Resultant Receiving Reliability			
		1 at 0.9	2 at 0.99	3 at 0.999	4 at 0.9999
		Reliability of Coders and Receivers Combined			
1	0.800	.720	.792	.7992	.7999
2	0.960	.864	.9504	.9590	
3	0.992	.893	.9821		
4	0.9984	.8986			

Phase 2: Transmitter, Coder, and Receiver System

Receivers	Coders	Combined Reliabilities (From Phase 1)	Number of Transmitters Used in Parallel and Resultant Transmitting Reliability			
			1 at 0.7	2 at 0.91	3 at 0.973	4 at 0.9919
			Reliability of Three-Functions System			
1	1	.720				.71
2	1	.792			.77	
3	1	.7992		.73		
4	1	.7999	.56			
1	2	.864			.84	
2	2	.9504		.86		
3	2	.959	.67			
1	3	.893		.81		
2	3	.9821	.69			
1	4	.8986	.63			

SOURCE: Adapted from a paper by Buddy L. Myers and Norbert L. Enrick, "Algorithmic Optimization of System Reliability" (Delivered at the 22nd Annual Technical Conference, American Society for Quality Control, Philadelphia, Pennsylvania, May 6, 1968).
NOTE: Blank spaces represent inapplicable combinations, since the optimum must represent a total of six units, and at least one coder, one receiver, and one transmitter are needed.

In phase 1, for example, the use of two coders with two receivers results in a combined reliability of 0.96 × 0.99 or 0.9504. This leaves two units for the transmitting function, since the total system is limited to six units.

From phase 2, we note that 2 transmitters, having a reliability of 0.91 jointly, combine with the prior reliabilities of 0.9504 to yield a system reliability of 0.9504 × 0.91 or 0.86.

All possible combinations are evaluated in this manner. Optimal reliability, it is apparent, occurs at the limit of six units for the system. Moreover, at least one of each of the three functions—coding, receiving, and transmitting—must be provided for. Blank spaces in the dynamic programming schedule represent combinations that are not feasible, according to the considerations just noted.

A search for the optimum is confined to phase 2, in which all three functions of the system are viewed in combination. Highest reliability, at 86 percent, occurs when two units of each of the three functions are utilized. Second best reliability, incidentally, results from a combination of one receiver, two coders, and three transmitters, at 84 percent.

Dynamic programming thus serves as a worthwhile design tool. Knowing the relevant component reliabilities and recognizing the time, space, and cost limitations involved in a project or system, the designer is enabled to plan redundancy in a manner that maximizes attainable reliability.

Concluding Observations

Created by the needs of the space age, the concept of reliability presents an important extension in the definition of product quality. Reliability has the virtue of being expressible in quantitative terms. Products can be tested for conformance to these requirements. Designs can be analyzed with a view to the components and arrangements needed to satisfy a particular end use.

Most of the applications of reliability work have come from military and space exploration activities. But the impact on commercial products has been noted. The life of appliances, the potency of pharmaceuticals, the stability of chemicals, and the performance of synthetic heart valves are all areas where reliability is of utmost importance. The next years will, undoubtedly, see vast extensions not only of the concepts and methods of reliability but also of the fields of application.

By its nature, reliability involves a probabilistic and thus uncertain type of prediction of performance characteristics. We can hope to reduce the risks of error in such predictions, but we cannot ever expect to attain certainty. Unavoidable as some allowance for error may have to be, the alternative of making no estimates, no evaluations, and no predictions is unthinkable in a modern society that is increasingly geared toward complexity of its appliances, equipment, and productive machinery.

REVIEW QUESTIONS

1. When components have been underdesigned, what problems can this produce?
2. What problems will arise from overdesigned components?
3. How is redundancy achieved?
4. A double-limit operation involves two failure possibilities. For example, if a switch may either fail to close or else close prematurely, we have a double-limit problem. Provision of redundancy tends to increase the probability of successful closure. Yet, it is not an unmixed blessing. Why?
5. In single-limit operation, where only one type of failure can occur, redundancy always increases reliability. Yet, even in the presence of unlimited production funds, there are practical limits that must be observed. What are these limits?
6. Time bases for component reliabilities forming part of a single system must all be comparable. In general terms, how is this comparability achieved.
7. Where and how are component reliability data obtained?
8. A three-element system consists of a search radar, a tracking radar and a missile launcher, with reliabilities of 0.9, 0.8 and 0.4 respectively. What is (A) present system reliability and (B) reliability if two missile launchers are used?
9. Four components are arranged in series, with the reliabilities and costs shown below:

Component Identification	A	B	C	D
Reliability, P_s	0.4	0.6	0.7	0.9
Cost, $	1	2	3	4

What are the reliabilities and costs for the following: (A) present system, (B) system with A tripled, (C) system with A tripled and B doubled? (Doubling and tripling are accomplished through parallel redundancy).

See Case 16, Appendix 3, for additional review material.

CHAPTER 19

Experiments Designed to Enhance Quality and Reliability

An age of expanding technological development brings with it the need for excellence in manufacturing. While there will be occasional breakthroughs in process and product innovations, the steadiest amount of progress overall accrues from laboratory and manufacturing floor experimentations, which seek step-by-step to enhance productivity and to economically achieve a highly marketable product in terms of cost, quality, and reliability. Typical questions to which experiments may seek answers are the following:

How can a product be redesigned to be more easily produced while yet able to perform more functions at higher reliability?

Which processing equipment is best for the firm's production and marketing needs?

How are production variables related to cost factors, and in what direction should modifications be made to obtain the ultimate from equipment?

Systematic investigations will not only provide valuable information for process and product improvement, but will also point out new areas of worthwhile research, leading to subsequent further gains. Over a period of time, multiple small improvements and innovations will thus grow to a major competitive advantage.

Experimentation can yield false and misleading information unless planned, executed, and analyzed in accord with the principles of scientific methods. Essentials of sound experimentation and data evaluation, illustrated with the aid of case histories, will be presented. Randomization, replication, and balance—the foundations of valid investigation—will be examined first.

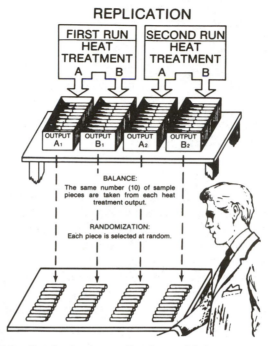

Fig. 19-1. Randomization, replication, and balance in experiments.

Randomization

In the running of experiments, as in quality control during production, information must come from randomly drawn samples. Random selection of individual sample units avoids possible bias. For the manufacturer considering two different heat treatments A and B, in Fig. 19-1, with outputs A_1, A_2, B_1, and B_2, it is important that each of the four samples of ten units be chosen randomly. In practice this means that the tester selects blindly from various parts of the box. A more formal approach, which is warranted in some situations, is to assign to each piece in a box a successive number, such as from 1 to 50, and then to choose each piece by reference to a table of random numbers.

Below are ten random two-digit numbers, taken as a block from a larger table of random numbers (see Table 19-1):

88 03 61 53 77 15 48 65 59 20 70 53
44 49 30 18 59 37 79 77 66 28 70 19

Table 19-1.   Random Sampling Numbers

Random Sampling Numbers

10	05	65	97	10	00	68	73	30	31	45	19	17	02	16	39	17	50	29	67
02	85	35	47	95	87	37	52	56	07	07	85	14	31	71	88	71	16	50	85
83	27	95	62	63	73	57	03	99	36	24	18	63	02	98	00	26	73	25	24
62	44	46	17	19	68	69	07	33	83	56	16	00	25	20	84	40	22	06	92
50	34	36	69	05	49	79	39	37	17	23	25	08	91	90	24	48	89	35	49
22	82	64	93	38	71	42	13	88	32	60	65	58	94	22	95	09	13	36	46
11	19	80	19	54	21	10	41	33	13	69	05	43	71	01	47	70	02	43	04
38	18	82	51	45	08	36	74	38	74	56	48	39	10	62	89	53	13	18	14
83	55	50	22	49	72	73	57	55	17	06	06	06	87	68	98	09	01	96	34
54	74	74	46	88	17	29	24	40	59	53	45	83	40	80	68	83	07	93	33
49	07	95	12	80	32	73	63	02	53	95	86	94	09	45	25	88	10	76	16
60	67	26	60	73	78	52	59	32	19	25	32	92	57	38	35	39	00	72	49
29	29	72	83	68	09	95	90	69	01	15	46	30	65	12	55	61	03	40	35
65	73	13	73	74	35	24	85	40	06	97	42	22	19	07	26	94	57	66	80
86	65	60	69	73	96	20	05	52	85	94	81	24	50	16	96	30	72	25	76
00	93	99	36	04	50	06	96	74	57	81	23	93	49	63	90	98	26	63	59
40	08	58	79	34	65	55	64	86	85	51	52	33	29	18	60	43	81	80	29
19	64	24	74	97	64	84	37	32	06	28	70	50	28	10	95	00	39	00	56
21	28	10	07	75	45	25	92	26	09	88	33	89	08	54	61	70	55	10	45
64	79	47	34	02	47	34	02	47	36	27	68	29	77	08	90	04	28	81	18
26	50	64	58	98	27	68	64	47	65	82	76	36	70	09	22	31	20	41	03
53	59	36	53	76	51	13	99	53	68	35	70	05	75	64	63	19	85	03	01
43	85	39	15	21	21	97	50	40	00	44	51	94	96	90	81	58	42	01	95
04	71	52	62	92	74	21	10	43	43	74	47	40	94	95	68	76	54	68	94
57	16	52	55	27	27	73	07	18	74	24	46	78	29	33	16	48	29	10	92
29	52	51	85	86	64	55	65	34	79	58	08	63	49	93	09	14	60	85	82
46	23	81	57	74	96	06	50	04	36	99	93	16	50	24	02	50	96	62	40
68	81	94	85	52	05	20	96	73	69	60	65	07	19	22	09	62	26	26	72
06	42	97	54	81	17	67	41	74	68	50	95	43	24	70	77	54	97	14	47
35	98	34	91	25	62	37	96	45	75	30	35	29	64	77	91	33	99	45	83
58	87	29	39	26	79	67	06	16	27	20	79	95	31	30	44	60	49	42	03
76	81	67	77	49	88	89	50	72	15	72	97	89	79	17	82	39	45	82	98
78	90	64	51	69	25	58	24	37	86	98	94	75	97	90	13	69	47	89	69
27	97	97	21	28	92	20	79	05	56	18	55	04	99	53	21	98	52	60	84
14	68	53	35	31	53	60	30	57	87	36	62	01	45	48	83	86	64	35	81
58	88	28	02	45	49	08	80	21	06	55	72	66	12	88	03	61	53	77	15
39	13	90	60	97	27	61	48	93	10	05	99	18	59	48	65	59	20	70	53
14	52	12	81	25	81	36	00	54	77	30	01	88	25	44	49	30	18	59	37
81	90	03	17	23	05	74	63	31	96	57	64	79	48	72	77	66	28	70	19
83	83	44	70	11	25	59	56	28	44	17	20	83	66	71	24	14	42	55	42

SOURCE: N. L. Enrick, *Industrial Engineering Manual for the Textile Industry*, 2nd ed., Huntington, New York: Robert E. Krieger Publ. Co.

Assume we have decided to proceed along each row (moving from column to column would have been the alternative), omitting inapplicable numbers (those above 50 and those that are duplications); we then find the following piece numbers to be drawn: 3, 15, 48, 20, 44, 49, 30, 18, 37, 28. This completes the sample of ten pieces for one of the four boxes.

The reason why random numbers are used is the reluctance of some samplers to draw, say, the first piece or to draw from the top layer. Others will emphasize corners of a box or special layers of a lot. It is thus difficult to argue with the contention that the best means of obtaining an unbiased sample is by means of random-number-dictated selection. It is, however, also the most time-consuming approach; and the instruction to the sampler to "just draw as randomly as you can" becomes a common substitute.

Replication

When an experiment is run more than once it is said to be *replicated*. Two runs, such as in Fig. 19-1, represent a replication r of 2. Had there been five runs, then one would say $r = 5$. Each time the experiment is run it involves an expenditure of time and money, and for this reason most replications encountered in practice do not exceed $r = 2$. Against the costs of extra runs must be weighed the benefits, as underlined in Fig. 19-2. These are:

1. The additional information provided by replication enhances the reliability of experimental outcomes.
2. From the repeat runs, a measure of the relative amount of experimental error, the *error standard deviation*, is gained.
3. By contrasting the differences among the averages of principal interest (that is, the experimental outcomes) against the limits of experimental error, it is possible to judge whether or not the outcomes are statistically significant (and thus not ascribable to chance fluctuations).

Experimental error represents the inherent variations in raw material, effects of vibration, temperature–humidity fluctuations, and other ambient conditions, as well as the inevitable variations in testing that are beyond the practical limits of control of a careful experimenter or experimenting team. No implication of "mistake" is attached to the term *error*.

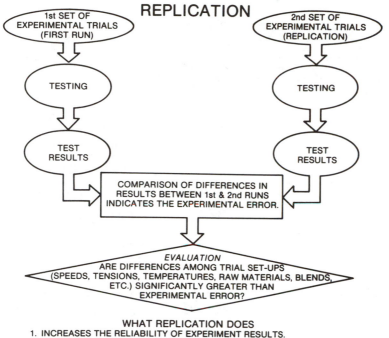

Fig. 19-2.   Nature and benefits of replication.

Balance

Each of the principal factors of an investigation must have an equal chance to affect the experimental outcomes. Thus in Fig. 19-1 balance requires that each of the four boxes be sampled equally. Beyond that it is important that the second run be in every regard equal to the first one. An unbalanced setup will yield distorted results, thus defeating the experimenter's effort to learn the truth.

A further example of the problem of balance in experimentation appears in Fig. 19-3. In studying the effect on quality of plastic versus metal tension posts, it would be erroneous to have all of one kind on the motor side of the winder, with all of the other kind on the bearing side. The coiling shaft is usually held firmer on the motor end, because of possible wear of the bearing. Then, when the product on the bearing side shows poorer quality, we do not know whether this is a true effect or whether it is the bearing effect. By inter-

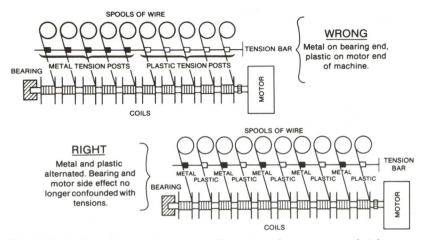

Fig. 19-3.   Balance in experimentation. Illustration shows wrong and right ways of design to achieve balance.

spersing the posts, as shown, any bearing wear that may exist will no longer be able to interfere with or "confound" the tension post findings.

Reliability Improvement Application

As a major illustrative example, the case of an investigation designed to discover means of producing a reliable electromechanical assembly will be presented. Reliability was measured in terms of "flexure-cycles to failure," and the following principal factors and levels were studied:

Factor	Three-Factor Replicated Study Levels at Which Factor Was Studied
1. Number of wire strands	Seven and nine strandings.
2. Wire slack	0, 0.03, 0.06, 0.09, and 0.12 in. of unsoldered, uninsulated wire to connection point.
3. Wire gage size	Numbers 24, 22, and 20.

The results appear in the separate tabulation headed "Wire Life Data With Estimate of Experimental Error." The error standard deviation is obtained from the average range R among the replications ($r = 2$ per cell) and the conversion factor $F_r = 0.8865$ from Table 6-2 (p. 78). Replications r in reexperimentation are analogous

Wire Life Data with Estimate of Experimental Error
(Each of the 60 Tests Represents Flexure Cycles, in 100s to Failure)

No. of Strands	Wire Slack[a] (inches × 10^{-2})	Wire Gage Size[b]										Grand Total of Ranges
		24			22			20				
		Run 1	Run 2	Range	Run 1	Run 2	Range	Run 1	Run 2	Range		
7	0	2	4	2	14	9	5	6	8	2		
7	3	5	2	3	6	15	9	5	7	2		
7	6	6	3	3	14	7	7	6	5	1		
7	9	9	16	7	12	12	0	8	12	4		
7	12	14	12	2	10	14	4	12	11	1		
9	0	3	3	0	10	14	4	12	11	1		
9	3	2	5	3	17	17	0	16	8	8		
9	6	5	5	0	10	10	0	10	8	2		
9	9	6	4	2	16	11	5	13	7	6		
9	12	13	15	2	20	17	3	12	15	3		
Total of Ranges				24			37			30		91

Experimental Error Standard Deviation

1. Using method based on ranges, obtain sum of ranges $R = 91$.
2. Since there are $k = 30$ ranges, average range $\bar{R} = 91/30 = 3.033$.
3. Factor F_r for converting $\bar{R}$ to estimated error standard deviation (from Table 5-1) is 0.8865. Therefore, error standard deviation σ is estimated to be $\bar{R}F_r = 3.033 \times 0.8865 = 2.7$.

[a] Slack represents the length of uninsulated, unsoldered wire at the connection point.

[b] Gage is a reciprocal measure; the larger the gage size, the smaller the wire diameter. Accordingly, sizes are shown in descending order.

SOURCE: Data reused with permission from N. L. Enrick, "An Analysis of Means in a Three-Way Factorial," *Journal of Quality Technology*, vol. 8, no. 4, 1976, © 1976 American Society for Quality Control.

Arithmetic Averages (Means) for Wire Life Experiment
(Each of the 30 Basic Entries Represents the Arithmetic Means
of Two Runs in 100s of Cycles to Failure)

No. of Strands	Wire Slack (inches × 10^{-2})	Wire Gage Size No.			Slack Mean	Group Mean	Grand Mean
		24	22	20			
7	0	3.0	11.5	7.0	7.2		
7	3	3.5	10.5	6.0	6.7		
7	6	4.5	10.5	5.5	6.8		
7	9	12.5	12.0	10.0	11.5		
7	12	13.0	12.0	11.5	12.2		
Strand Mean		7.3	11.3	8.0		8.9	
9	0	3.0	12.0	11.5	8.8		
9	3	3.5	17.0	12.0	10.8		
9	6	5.0	10.0	9.0	8.0		
9	9	5.0	13.5	10.0	9.5		
9	12	14.0	18.5	13.5	15.3		
Strand Mean		6.1	14.2	11.2		10.5	
7 & 9	0	3.00	11.75	9.25	8.0		
7 & 9	3	3.50	13.75	9.00	8.8		
7 & 9	6	4.75	10.25	7.25	7.4		
7 & 9	9	8.75	12.75	10.00	10.5		
7 & 9	12	13.50	15.25	12.50	13.8		
Group Mean		6.7	12.75	9.60			9.7

NOTES: 1. Each principal entry represents the mean of two runs. Thus, for 7 strands, zero slack, and wire size 24, the entry 3 = (2 + 4)/2.
2. Group means are obtained by averaging of entries. For example, strand means 7.3, 11.3, and 8.0 yield the group mean 8.9, representing 7 strands averaged over all wire gage sizes.
3. The grand mean 9.7 is the average of the combined (or group) means, and may also be obtained by averaging all 30 individual entries.

to sample size n in process sampling. The intersections of the two strands, five wire slacks, and three gages form $2 \times 5 \times 3 = 30$ cells. The cell means are shown separately in the tabulation headed "Arithmetic Averages (Means) for Wire Life Experiment."

Degrees of Freedom, *DF*

In ordinary sampling, given a sample size n, $DF = n - 1$. Applying this principle to experiments, each cell has $r - 1$ degrees of freedom,

DF. For $k = 30$ cells then, $DF = k(r-1) = 30(r-1)$. When $r = 2$, as in our (and most) applications, $DF = 30(2-1) = 30$. Had there been $r = 4$, then DF would be $30(4-1) = 90$. The concept of DF is important for entering the table of h_d factors, as discussed next.

Decision Lines

Comparable to control limits for the evaluation of sample averages in hourly or daily quality control, for experiments we need criteria for judging when a particular average is significantly different from the grand mean of the data. Developed by Ott,[1] they are known as decision lines, DL. In Tables 19-2 and 19-3, factors h_d for calculating DL are given for 95 and 90 percent confidence levels (involving corresponding risks of error of 5 and 10 percent). The calculations are

$$DL = \text{Grand Mean} \pm h_d \text{ (Error Standard Deviation)}/\sqrt{N},$$

where N is the total number of observations in the experiment. For our example, with grand mean of 9.7, error standard deviation of 2.7, and $N = 60$, we have for a 95 percent confidence level

$$DL = 9.7 \pm 18.58(2.7)/\sqrt{60}$$
$$= 9.7 \pm 6.5$$
$$= 3.2 \text{ to } 16.2$$

in hundreds of cycles to failure. The value $h_d = 18.5$ is obtained from Table 19-2 at the level of $DF = 30$ and $k = 30$.

Had we chosen a more liberal 90 percent confidence level, h_d would have been 17.2, and then

$$DL = 9.7 \pm 17.2(2.7)/\sqrt{60}$$
$$= 9.7 \pm 6.0$$
$$= 3.7 \text{ to } 15.7$$

or from 3,700 to 15,700 flexure cycles. These narrower limits will admit more experimental averages as significant, but they also incur a higher risk of erroneously calling an effect "significant," when actually it is merely the effect of chance fluctuations (experiment error). The risk, as noted, is $100 - 90$ percent, or 10 percent.

---

[1] E. R. Ott, "Analysis of Means—A Graphical Procedure," *Industrial Quality Control,* vol. 24, no. 2, pp. 101–109, 1967.

Table 19-2. Factors h_d for 95 Percent Confidence Level Decision Lines
(Number k of Means Under Comparison)

Degrees Freedom, D.F.	2	3	4	5	6	7	8	9	10	11	12	14	16	18	20	24	30	40	60
2	4.30	6.73																	
3	3.18	5.60																	
4	2.78	4.99	7.47																
5	2.57	4.65	6.60	8.06															
6	2.45		6.10	7.42	8.63														
7	2.36	4.43	5.79	7.00	8.14	9.01													
8	2.31	4.27	5.56	6.72	7.78	8.70	9.74												
9	2.26	4.14	5.39	6.50	7.51	8.45	9.39	10.24											
10	2.23	4.06	5.27	6.24	7.34	8.25	9.13	9.96	10.74										
11	2.20	3.99	5.16	6.22	7.18	8.06	8.92	9.73	10.50	11.23									
12	2.18	3.93	5.08	6.10	7.04	7.94	8.78	9.53	10.29	11.00	11.71								
13	2.16	3.89	5.02	6.02	6.95	7.81	8.63	9.39	10.11	10.81	11.51								
14	2.15	3.85	4.95	5.96	6.86	7.72	8.49	9.25	9.99	10.66	11.34	12.58							
15	2.13	3.80	4.92	5.90	6.80	7.62	8.41	9.14	9.87	10.53	11.18	12.44							
16	7.12	3.76	4.87	5.84	6.73	7.54	8.33	9.05	9.85	10.44	11.08	12.29	13.44						
17	2.11	3.75	4.83	5.80	6.66	7.50	8.25	8.97	9.66	10.34	10.98	12.19	13.32						
18	2.10	3.73	4.80	5.76	6.62	7.42	8.18	8.91	9.60	10.24	10.87	12.08	13.21	14.27					
19	2.09	3.72	4.78	5.72	6.57	7.37	8.12	8.85	9.51	10.18	10.77	11.97	13.09	14.27					
20	2.09	3.69	4.76	5.70	6.55	7.35	8.10	8.80	9.45	10.12	10.71	11.90	13.01	14.05	15.52				
24	2.06	3.63	4.68	5.60	6.44	7.20	7.94	8.63	9.27	9.90	10.51	11.65	12.70	13.73	14.73	16.55			
30	2.04	3.59	4.61	5.50	6.31	7.08	7.78	8.46	9.09	9.71	10.28	11.39	12.43	13.44	14.38	16.16	18.58		
40	2.02	3.54	4.54	5.40	6.22	6.96	7.65	8.29	8.91	9.52	10.08	11.18	12.19	13.44	14.08	15.78	18.15	21.67	
60	2.00	3.48	4.47	5.32	6.10	6.83	7.49	8.14	8.73	9.33	9.88	10.92	12.08	12.86	13.77	15.44	17.72	21.17	27.04
120	1.98	3.44	4.39	5.24	5.99	6.71	7.36	7.98	8.58	9.14	9.68	10.71	11.93	12.58	13.47	15.11	17.34	20.67	26.35
Inf.	1.96	3.38	4.33	5.16	5.90	6.59	7.22	7.83	8.43	8.98	9.52	10.49	11.46	12.33	13.16	14.77	16.91	20.17	25.66

SOURCE: Based on tables and relations developed in Nelson, Lloyd S. "Factors for the Analysis of Means," *J. of Quality Technology*, vol. 6, no. 4, pp. 175–181, Oct. 1974.

Table 19-3. Factors h_d for 90 Percent Confidence Level Decision Lines
(Number k of Means Under Comparison)

Degrees Freedom, D.F.	2	3	4	5	6	7	8	9	10	11	12	14	16	18	20	24	30	40	60
2	2.35																		
3	2.13	5.29																	
4	2.02	4.51	6.06																
5	1.94	3.89	5.47	6.72															
6	1.90	3.89	5.14	6.28	7.36														
7	1.86	3.73	4.92	6.00	7.00	7.94													
8	1.83	3.63	4.76	5.80	6.75	7.64	8.49												
9	1.81	3.55	4.66	5.64	6.55	7.42	8.23	8.99											
10	1.80	3.49	4.56	5.52	6.42	7.25	8.04	8.80	9.51										
11	1.78	3.44	4.49	5.44	6.30	7.13	7.88	8.63	9.33	9.99									
12	1.77	3.39	4.43	5.36	6.22	7.01	7.75	8.49	9.15	9.83	10.4								
13	1.76	3.37	4.38	5.30	6.15	6.93	7.67	8.37	9.03	9.68	10.3								
14	1.75	3.34	4.35	5.24	6.08	6.86	7.57	8.26	8.94	9.56	10.2	11.4							
15	1.75	3.31	4.31	5.20	6.02	6.79	7.51	8.17	8.85	9.46	10.1	11.2							
16	1.74	3.30	4.28	5.16	5.97	6.74	7.43	8.12	8.76	9.39	9.98	11.1	12.2						
17	1.73	3.28	4.26	5.14	5.93	6.69	7.38	8.06	8.70	9.30	9.88	11.0	12.1						
18	1.72	3.25	4.24	5.10	5.90	6.64	7.33	8.00	8.64	9.23	9.82	10.9	12.0	13.0					
19	1.72	3.24	4.21	5.08	5.88	6.61	7.30	7.95	8.58	9.17	9.75	10.9	11.9	12.9					
20	1.71	3.24	4.19	5.06	5.84	6.56	7.25	7.92	8.55	9.14	9.72	10.8	11.9	12.8	13.7				
24	1.70	3.20	4.14	4.98	5.75	6.47	7.14	7.78	8.40	8.98	9.55	10.6	11.6	12.6	13.5	15.2			
30	1.68	3.15	4.09	4.92	5.68	6.37	7.01	7.67	8.25	8.82	9.35	10.4	11.4	12.3	13.2	14.9	17.2		
40	1.67	3.11	4.04	4.84	5.59	6.27	6.93	7.52	8.10	8.66	9.22	10.2	11.2	12.1	12.9	14.6	16.8	20.2	
60	1.66	3.08	3.98	4.78	5.50	6.17	6.83	7.41	7.98	8.54	9.05	10.1	11.0	11.9	12.7	14.3	16.5	19.7	25.3
120	1.65	3.04	3.93	4.72	5.43	6.10	6.72	7.30	7.86	8.38	8.89	9.88	10.8	11.6	12.5	14.0	16.2	19.3	24.7
∞	1.64	3.01	3.88	4.66	5.34	6.00	6.61	7.18	7.74	8.25	8.76	9.70	10.6	11.4	12.2	13.8	15.8	18.9	24.1

SOURCE: Based on tables and relations developed in Nelson, Lloyd S. "Factors for the Analysis of Means," *J. of Quality Technology*, vol. 6, no. 4, pp. 175–181, Oct. 1974.

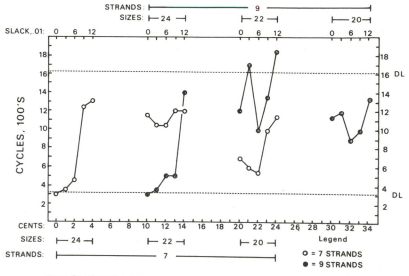

Fig. 19-4. Multicomparative decision line graph. The effects of strands, sizes, and slacks are related to product flexure life (in 100s of cycles to failure). In addition a scale of relative costs is presented. Points outside *DL* show significantly better or poorer performance.

Multicomparative Graphic Analysis

Knowing the 30 cell means and their *DL*s is not enough. Now it is essential to plot the result of the three-factor replicated study in a way that the important findings can be seen in readily comparable overview. While opinions may vary on the best format, a highly effective means of presentation is the approach shown in Fig. 19-4, showing the contribution to flexure life of (1) strands, (2) slacks, and (3) sizes. In addition one can usually manage to introduce a scale of relative costs (as is done in our example). Five variables thus appear in uncluttered, readily contrastable form, on one graph. Moreover, the *DL*s show the two instances of significantly superior performance: strands = 9, size = 22, and slacks = 0.03 or 0.12.

Based on this information the decision makers can now (1) select the most suitable combination for future full-scale production or (2) arrange for additional experiments (such as may have suggested themselves from the results of the present investigation).

Factor averages may also be contrasted, such as for the five slacks, the three sizes, and the two strands in Fig. 19-5. Again, 0.12-in.

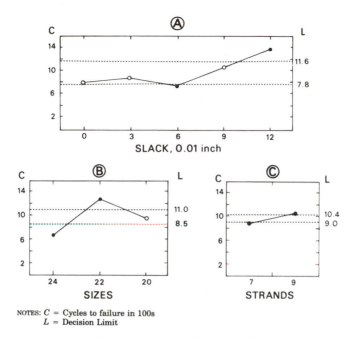

NOTES: C = Cycles to failure in 100s
L = Decision Limit

Fig. 19-5.   Comparison of group averages. This information is easier to comprehend than the presentation of individual interaction effects, but it lacks the detail needed by the researcher seeking maximum information from the experiment.

slack, size 22, and 9 strands point to highest reliability; but detail is lost, such as the good results obtained with the somewhat less expensive 0.03-in. slack (at identical strand and size).

Finally, one may also plot three sets of two-way effect means, as in Fig. 19-6. These are for the effect on product life, in terms of flexure cycles to failure, of (1) strands and slacks averaged over size, (2) sizes and slacks averaged over strands, and (3) sizes and strands averaged over slacks. This shows the same optimal value as the prior averages, plus a new entry: the less expensive size 24 at 0.12-in. slack. It should be clear, however, that this type of graph is quite difficult to evaluate. In general, it is best to use charts of group averages merely as supplements to the individual cell means.

Further *DL* Calculations

Determination of the additional *DL*s for the charts for group averages proceeds in the manner described previously, recalling the

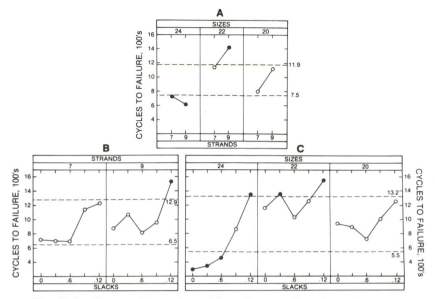

Fig. 19-6.   Two-way interactions of three-factor replicated experiment.

grand mean of 9.7, with an error standard deviation of 2.7 and $N = 60$. Since the number k of averages under comparison for slacks, sizes, and strands, respectively, is 5, 3, and 2:

$$DL \text{ for Slacks} = 9.7 \pm 5.5(2.7)/\sqrt{60}$$
$$= 9.7 \pm 1.9$$
$$= 7.8 \text{ to } 11.6;$$
$$DL \text{ for Sizes} = 9.7 \pm 3.59(2.7)/\sqrt{60}$$
$$= 9.7 \pm 1.25$$
$$= 8.45 \text{ to } 10.95$$
$$= 8.5 \text{ to } 11.0 \text{ rounded};$$
$$DL \text{ for Strands} = 9.7 \pm 2.04(2.7)/\sqrt{60}$$
$$= 9.7 \pm 0.7$$
$$= 9.0 \text{ to } 10.4.$$

It should be sufficient to illustrate just one calculation for the two-way means. For the effect of strands and slacks, averaged over sizes, involving ten strand–slack combination averages, $k = 10$ and hence $h_d = 9.09$. Next,

$$DL \text{ for Strand–Slack Effects} = 9.7 \pm 9.09(2.7)/\sqrt{60}$$
$$= 9.7 \pm 3.2$$
$$= 6.5 \text{ to } 12.9$$

in terms of 100s of flexure cycles to failure.

Relation of *DL* to Control Limits

Decision lines DL and control limits CL have in common the use of a band within which variations are ascribed to chance fluctuations and beyond which they are viewed as significant. Both use essentially the smallest applicable standard deviation to calculate limits; that is, the error standard deviation is the counterpart of the within-sample standard deviation. There are these differences:

1. DLs are based on a limited number k of cells, while CLs are usually based on a larger series of samples.
2. Control limits are calculated with simple reference to the normal curve, while for DLs the factors h_d make an allowance for less data from which the standard deviation is estimated.

Historically, CLs preceded DLs. DLs were developed by Ott,[2] who wanted for the evaluation of results from experiments the same types of graphic tools that serve for quality control during manufacturing.

For the DL approach to work it is generally assumed that there is so-called "homoscedasticity," that is, that the individual cell variances have similar standard deviations. One way to test for homoscedasticity is to set up a control chart for ranges, as in Fig. 19-7. Since all of the points for the 30 individual ranges fall below the CL, homoscedasticity is demonstrated.

The range control limit of 9.9 in flexure cycles to failure (in 100s) is obtained as prescribed in Chapter 5 on *Control of Variability*. In detail, using $\overline{R} = 3.033$ and factor D_4 for $n = r = 2$ of 3.267, $CL = 3.033 \times 3.267 = 9.9$ as the upper limit. No lower limit is needed.

Interaction

The concept of interaction is best explained with the aid of our illustrative case of flexure life of an electromechanical device. In Fig. 19-4 is shown the configuration of the 30 individual cell averages, each reflecting the interaction of a particular level of size, slack, and

---

[2]E. R. Ott, *ibid.*

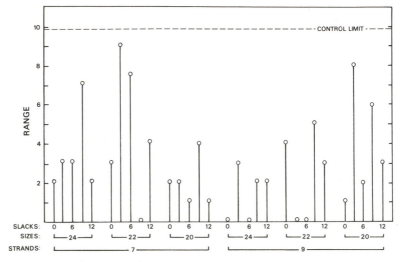

Fig. 19-7.   Control chart for within-cell ranges.

strand. These so-called *three-way interactions* are considered statistically significant when one or more points fall outside control, that is, outside the decision limits.

Two-way interactions, in Fig. 19-6, are again significant, with two-way averages beyond *DL*. Finally, there cannot be any interactions in the means of Fig. 19-5, since the final group averages only appear. We call these the *main effects*.

Absence of Interaction

When individual cell averages occur in approximately parallel lines (or curves), such as in Fig. 19-8, one generally notes that there are no significant interaction effects. The data refer to the effect on the strength of a ceramic electric insulator of two contours (A_1 and A_2), three heat-treating cycles (C_1, C_2, and C_3), and four glazes (B_1 to B_4). No point falls outside *DL*. Yet, while the interactions are not significant, one main effect is. Figure 19-9 shows contour A_2 significantly superior.

Absence of interaction makes data easier to review, but the real opportunities for cost savings occur in the presence of the special configurations observed among two- and three-way interactions. For example, the significant two-way-interaction effect (in Fig. 19-6) of

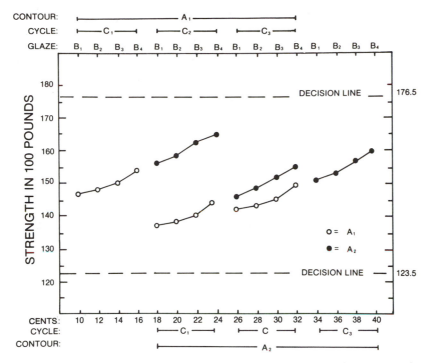

Fig. 19-8.   Individual cell averages of experimental investigation of contours, cycles, and glazes on ceramic insulator strength. Decision lines and costs are included.

0.12-in. slack and 24-gage wire suggests that this relatively inexpensive combination may be a viable alternative for the more costly 0.12-in. slack with 22-gage, at 9 strands.

Interaction—Additional Example

It is clear that the concept of interaction is important in all experimentation; and it will be useful to demonstrate an additional case in Fig. 19-10, reflecting the effect of forming pressure (in pounds-per-square inch, psi) on processing rejects in percent. The optimal (lowest rejects) point for product A is 120 psi, but for product B it is 130 psi. By knowing this differential effect or, in other words, this *interaction* effect, we are able to establish specific processing standards that assure the lowest rejects for each product. In practice there may be 6 to 20 products, and for each a separate search for the

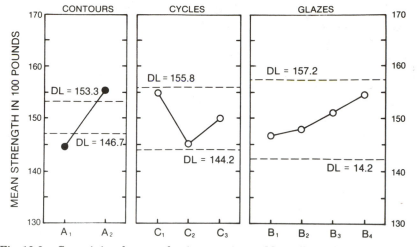

Fig. 19-9.   Ceramic insulator production experiment. Main effects of contours, cycles, and glazes are compared.

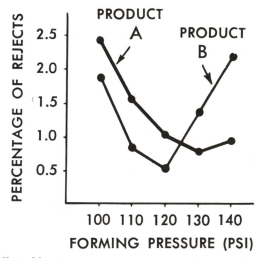

Fig. 19-10.   Effect of forming pressure on reject rate. As a result of pressure–product interaction, the optimum for product A is 120 psi, while for B it is 130 psi. Interaction may thus be viewed as having produced "differential" effects.

optimum would seem warranted in terms of the ultimate cost savings.

On the other hand, a person unaware of interaction effects might have merely sought an average curve for the two (or more) products. This curve would have bottomed at approximately 115 psi, the "average optimum" but not the overall lowest when individual product needs are considered.

Those who have engaged in a good deal of industrial experimentation are aware of the fact that in most situations the experiments worth running are the ones that involve interactions. The challenge then is to identify those points in the configuration of experimental outcomes that best suit the quality and cost needs of the production system.

Economic Use of Interaction Knowledge

An analysis of the way in which interaction effects occur can help a manufacturer save costs and enhance productivity. In order to demonstrate this fact Fig. 19-11 has been provided. The right-hand side shows a *theoretical* interaction-free case, while the left-hand side shows the *actual* interaction that takes place.

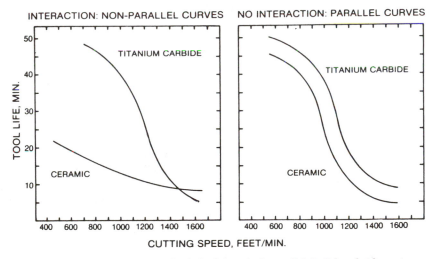

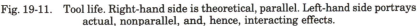

Fig. 19-11.   Tool life. Right-hand side is theoretical, parallel. Left-hand side portrays actual, nonparallel, and, hence, interacting effects.

SOURCE: Norbert L. Enrick, *Management Handbook of Decision-Oriented Statistics*, Melbourne, FL: Krieger Publ. Co., 1980.

Now, if the right-hand side were true, life would be relatively simple. A general statement could be made for the life of titanium carbide versus ceramic tools: "at all practical speeds, titanium is 10 percent longer lasting than ceramic as a cutting tool." Humans love wide-ranging generalizations, but, unfortunately, they are usually not true.

The left-hand side shows the actual situation. It says this: "at slow speeds, titanium carbide is the superior life tool, but at very high speeds, ceramic is better." Moreover, the graph shows the precise point when this cross-over occurs, which is a cutting speed of 1500 feet per minute.

With this information at hand, we can now look at productivity, labor, and materials costs at various speeds and with alternate tools, arriving at an optimum as regards costs and productivity.

Thus interaction effects are a nuisance in that they do not permit us to generalize. They are a boon in providing precise information to the astute manufacturer on how to attain optimal results under varying conditions of market need and manufacturing resources and facilities available.

Experimentation during Production

Increasingly, a two-fold trend in experimentation accompanies the development of high-precision, high-technology, and biochemically advanced products:

1. As one study investigates phenomena and uncovers vital information, it becomes clear that further unexplored areas need study. Often these areas had not been recognized until research was done.
2. Competition, world-wide, is intense. Thus there is the quest to arrive in the market place "first" and with "the most." This urgency factor, in turn, means that production will often start before products and processes have been finalized. Modifications and improvements occur as production proceeds.

It should be emphasized, of course, that rushing into production without the utmost of preparation effort can be very dangerous. Certainly, it is possible to leave some detail to be worked out during production, but the essentials of manufacturing must have been developed.

As a case in point, reference is made to the original production of penicillin. The pharmaceutical firm rushed into production without all detail having been worked out (see Fig. 19-12), but the essentials had been researched and researched well: (1) how to do deep-tank fermentation, and (2) which mold, from an extensive search, would be the most efficient strain.

No good formula can be given on how much laboratory research, how much design and redesign effort, and how much computer simulation and related research should precede the actual processing stage. Nor are there any good guidelines how to proceed in gearing up, from small scale to larger and larger runs. Management and engineering–technological knowledge and judgment are involved.

Production Experimentation—The Case of Penicillin

Background
In the late 1930s, the value of penicillin became known through Dr. Alexander Fleming's research. However, production in surface fermentation flasks produced only minute quantities that were extremely difficult to extract and impure.

Brainstorming
Pfizer Laboratories was among the pharmaceutical firms interested in producing penicillin. Brainstorming investigation of practical means of production hit on an idea that had worked previously in gluconic acid production: deep-tank fermentation.

Laboratory Research
Laboratory research was needed to find a penicillin strain that would thrive in deep tank fermentation. *Penicillium chrysogenum* emerged, with 200 times the yield of the original strains tried.

Production
Time compression was of the essence. Production experimentation with 50-gallon fermentation tanks was initiated in search of:

1. Efficient extraction and recovery of penicillin.
2. Removal of impurities.
3. Safe, stability-promoting drying and packaging.
4. Quick biochemical assay of potency, stability, and purity.

Production was increased in stages until it reached 10,000-gallon tanks. Throughout, yield, recovery, purity, potency and stability were tested and assured.

By-Products

1. Knowledge, skill, and ability in the antibiotics manufacturing field.

2. Leadership position in antibiotics production with a reputation for timely, high-quality, and versatile response to health services needs.

Fig. 19-12.   Production experimentation, a case.

Moreover, there are risks either way. Slow and careful proceeding may be safe product-cost wise, but may arrive at the market place late. Accelerated innovation will ensure timely arrival, but hidden flaws may be discovered later, and cause long-range difficulties. These types of problems are, of course, what business is all about.

Error Standard Deviation

In our principal example, experimentation was done in search of the best reliability of an electromechanical device. A replication r of 2 was used. This number of replications is used in most experiments in industry. While a greater r would enhance the precision of the estimate of the experimental error standard deviation, it would also in many cases add unacceptable costs and complexities.

For simplicity, we showed the estimation of the error standard deviation from the average range $\overline{R}$. Naturally, a somewhat more precise method would rely on an average of the variances for each of the 30 cells. For the first cell, for example, with individual observations of 2 and 4 and a mean of 3, we would find:

Individual Observation	Mean	Deviation from Mean	Square Deviation
2	3	-1	1
4	3	$+1$	1
Sums		0	2

The degrees of freedom, DF, is $n - 1 = 2 - 1 = 1$. Hence for this cell the estimated variance is

$$\sigma^2_{\text{cell}} = \Sigma \text{ squared deviations}/DF$$
$$= 2/1$$
$$= 2$$

Proceeding similarly for all 30 cells, a total of the 30 variance estimates of 226.5 is obtained. Hence,

$$\sigma^2_{\text{experiment}} = \Sigma \sigma^2_{\text{cells}}/k$$
$$= 226.5/30$$
$$= 7.55$$

and, taking the square root,

$$\sigma_{\text{experiment}} = 2.748$$

which is close to the value obtained from the average range.

Error Standard Deviation—Special Case

In those instances where $r = 2$, a simplification in estimating the error variance is possible. In particular, the reader will note that the value $\sigma^2_{cell} = 2$ is comparable to taking the original range, $R = 2$, squaring it, and then dividing by 2:

$$\sigma^2_{cell} = R^2/2$$
$$= 2^2/2$$
$$= 2$$

For the next cell, the range is 3, and therefore,

$$\sigma^2_{cell} = 3^2/2$$
$$= 9/2$$
$$= 4.5$$

while the more tedious approach based on individual observations would again give an identical result:

Individual Observation	Mean	Deviation from Mean	Squared Deviation
5	3.5	+1.5	2.25
2	3.5	−1.5	2.25
Sums		0	4.5

so that

$$\sigma^2_{cell} = 4.5/1$$
$$= 4.5$$

which corresponds to the square of the range divided by 2. Generalizing, we may state:

$$\sigma^2_{experiment} = \Sigma R^2/N$$
$$= 453/60$$
$$= 7.55$$

as before. Here N is the total number of observation, which is $30 \times r = 30 \times 2 = 60$. Again, this simplification applies only when $r = 2$.

Normality of Error Term

Decision line analysis, as most analyses of experiments, relies on the assumption that the error variance is normally distributed. Usually, this assumption is well justified. If doubts occur, one can plot the deviation from mean data, as illustrated above, on normal probability paper. A normal distribution should result in a straight-line curve. Unfortunately, when normality is not fulfilled, treatment of the data can become quite complex, requiring attention by a specialist with considerable educational and experience background.

Concluding Observations

Experimentation in the laboratory, the pilot plant, or the production environment serves as a means of searching for conditions that will enhance and possibly optimize the factors of quality, cost, reliability, and ultimate marketability of product. Where these activities are pursued on a systematic, scientific, and well-planned basis, they represent an important mark of a modern, forward-looking, and competitive manufacturing group.

What we have presented are the essential principles and methods of sound, information-harvesting experimentation, followed by meaningful analysis. Furthermore, we have stressed simplified presentation, emphasizing graphic portrayal of the findings, contrasted against the decision lines. Visual, multicomparative contrasting of the effects of main factors and interactions among the experimental data facilitates the overview and ready comprehension of results. Management and staff people concerned are thus aided in the task of drawing conclusions from the investigation and arriving at decisions for implementation, to translate trial outcomes into routine realities in production.

REVIEW QUESTIONS

1. A manufacturer obtains raw stock from two suppliers, A and B, and then processes the material in one of two pieces of equipment, C or D. A check over two months shows the following percent defective results:

		Supplier A	Supplier B
Equipment C	March	2.0	3.1
	April	2.2	3.0
Equipment D	March	3.8	2.5
	April	3.7	2.5

Find the error standard deviation, based on the assumption that the two months represent replications.

2. Plot the data in Question 1, and calculate and add decision limits.
3. Is there evidence of significant interaction in the plotted data? Explain.
4. It has been said that significant interaction represents special burdens of presentation and evaluation for the analyst, but it also shows management special opportunities. Can you illustrate this?
5. For the principal illustration of this chapter, on the effect of strands, slacks, and wire gage size on flexure life, calculate all decision lines on the basis of 90 percent confidence limits. (The text presents the case of 95 percent.)
6. In connection with the analysis of experimental data, what is the purpose of the control chart for within-cell ranges?
7. It has been said that in-plant experimentation is rarely a one-time project. What, in effect, will occur?
8. A manufacturer of miniaturized valves, with wall sections so thin that problems of undue porosity occurred, wished to reduce the frequency of "leakers." It was clear that existing production methods were inadequate, and an experiment was run involving four pouring temperatures and two dwell times, with the results in leakers per 100 valves shown below:

Dwell times	Replicate	Pouring Temperatures Applied			
		High	Medium I	Medium II	Low
Long	1	2.2	1.0	2.5	1.8
	2	2.0	0.5	3.0	0.5
	3	1.8	0.0	2.0	0.7
Short	1	2.0	2.0	3.5	1.5
	2	1.5	2.5	3.4	2.0
	3	1.0	3.0	3.6	2.5

Find the error standard deviation and decision lines. Then, plot the results and discuss your findings.

CHAPTER 20

Quality and Reliability Experiments, Additional Applications

Up to this point we have discussed design and analysis of production and laboratory experiments involving two principal features:

1. The observations taken were variables, that is, they represented measurements along a scale.
2. Replication was used to obtain an estimate of the experimental error variation.

Situations may occur, however, where either one or both of these features are missing. For example, instead of variables we may be dealing with proportions or rates, and no replications may have been made. The resultant conditions may make it somewhat more difficult to establish significance of difference among experimental outcomes, but valid analyses are nevertheless quite readily obtainable.

Attributes Data

So-called attributes or go–nogo data represent test results recorded in only one out of two possible categories, such as good versus bad, or acceptable versus unacceptable. Resultant proportions, such as "percent effective" and "percent defective" will add up to 100 percent. For example, if the percent defective is 10, then the percent effective must be 90. We may also say that $p = 0.1$ and $q = 0.9$ in terms of decimal equivalents, where p and q represent "fraction defective" and "effective," respectively.

Snap action thermostat life under four competing alloy compositions was under study. For each of the four compositions or "treatments," 144 units were tested simultaneously on a test bank simulating 10,000 cycles of on-and-off demand, corresponding to the likely practical operating requirements during the years of use of the thermostat. The results appear below.

	Alloy Composition				
	A	*B*	*C*	*D*	Average
Number of units under test	144	144	144	144	
Number of units failing	12	8	1	24	
Fraction failing	0.083	0.055	0.007	0.167	0.078
Percent failing	8.3	5.5	0.7	16.7	7.8

The error standard deviation for this experiment is now found from our familiar formula:

$$\sigma_p = \sqrt{pq/n}$$
$$= \sqrt{0.078(0.922)/144}$$

where 0.922 is q or $1 - 0.078$. From Table 20-1 of factors h_p for decision lines *DL*, using a 95 percent confidence level as an example, and noting that k, the number of treatments, is 4, we find $h_p = 2.16$. Hence,

$$DL = p \pm h_p\sigma_p$$
$$= 0.078 \pm 2.16(0.027)$$
$$= 0.078 \pm 0.058$$
$$= 0.02 \text{ to } 0.136$$

or 2 to 13.6 percent. Alloy *C* has a significantly better reliability while Alloy *D* is inferior in relation to the average reliability of the set.

Factor h_p

It was previously observed that *DL*'s are obtained by methods similar to those for control charts, except that the factors for multiplying the standard deviation (or standard error), which are 2.0 and 3.0 for 95 and 99.7 percent confidence level control limits, respectively, are replaced by factors h for *DL*'s. These factors h vary with k, the number of treatments under comparison. While factors h are, like ordinary control limit factors, based on the normal curve, an adjustment is required to allow for k because of these reasons:

1. For control chart purposes, individual decisions occur *one at a time* at a particular point in time. An observed average or

Table 20-1.   Factors h_p for Decision Lines for Attributes and Rates

Number k of Proportions under Comparison	Confidence Level, Percent		
	90	95	99
2	1.16	1.39	1.86
3	1.74	1.95	2.40
4	1.94	2.16	2.62
5	2.08	2.31	2.76
6	2.18	2.41	2.87
7	2.27	2.49	2.95
8	2.34	2.55	3.02
9	2.39	2.61	3.07
10	2.45	2.67	3.12
15	2.62	2.83	3.29
20	2.73	2.94	3.31
30	2.88	3.09	3.53
60	3.11	3.31	3.73

Example: One hundred insulators per each of three different designs were subjected to accelerated stress tests, with proportion failing of 0.2, 0.35, and 0.05, so that $\bar{p}$ = 0.60/3 = 0.20. The standard error is $\sqrt{0.20(1 - 0.20)/100}$ = 0.04. To find 90 percent confidence decision limits,

$$
\begin{aligned}
DL &= \bar{p} \pm h_p \sigma_p \\
&= 0.2 \pm 1.94(0.04) \\
&= 0.2 \pm 0.0776 \\
&= 0.1224 \text{ to } 0.2776 \\
&= 12.24 \text{ to } 27.76 \text{ percent.}
\end{aligned}
$$

Therefore, the third design with p = 0.05 = 5 percent represents a significant improvement in performance characteristics. We note also that p = 0.35 is significantly high, compared to the DL of 0.2276.

SOURCE: Factors h_d divided by $\sqrt{k}$. For attributes testing, $DF = \infty$.

proportion must be evaluated as to whether or not the sampling result indicates an "in-control" or "out-of-control" condition in the production process.

2. For experiments, interest centers on all individual treatment results *simultaneously*. In our illustration, all four proportions were reviewed and contrasted together, identifying the significantly better and significantly poorer values. It is clear that DL's, designed for such simultaneous evaluations, require an appropriate adjustment, as reflected in h, which in turn corresponds to k.

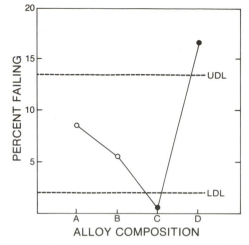

Fig. 20-1. Analysis of attributes data. Comparison of percent failures against lower and upper *DL*'s (*LDL* and *UDL*) shows alloy *C* to be best, while alloy *D* is poorest.

3. A further adjustment is needed for the factor h for DL, since usually an experiment involves a relatively small number of observations on which the estimate of error variability is obtained. By contrast, control charts demand considerable past experience data regarding the variability of a manufacturing process.

Much erudite work has been published regarding these special problems, which are beyond the scope of a text dealing with the principles and practice of applications of quality control and reliability assurance.

Rates of Occurrence

Observations need not be in terms of variables or proportions, but can instead represent a rate of occurrence. Some examples are:

Number of shipping errors per month.

Number of customer complaints per week.

Number of accidents per 100,000 miles of a trucking fleet.

Number of defects per 1000 yards of a fabric, such as filling bars, floats, slubs, spots, and torn selvedge.

Some of these errors could be converted to percent. For example, if track is kept of the number of shipments per month, the percent of shipping errors could be found. Others are impossible to convert. For example, the Poisson distribution expressing rates was first applied to the number of Prussian officers killed per year by the kick of a horse. To convert this rate to a percent, we would have to know the "opportunity area" involved, that is, the number of exposures of officers per year—information that, of course, is unobtainable. The Poisson distribution continues to be called the "horsekick distribution" in some references.

Many other situations in which rates are the only viable form of expression for data are encountered regularly: the number of lightning strokes per thunderstorm, the number of microorganisms per square millimeter of a microscope slide, the white cell count per cc of serum, or the number of fire alarms per year.

Decision Lines for Rates

Prior to weaving, a mill will apply sizing to the warp beams, for the purpose of (1) imparting body to the fabric and (2) reducing loom stops due to warp breaks. A weaver experimented with three formulas, A, B, and C, over a 4-week period:

	Sizing Formula			
	A	B	C	Average
Warp stops over 4 weeks	28	16	7	17

The average rate, $\bar{c}$, is 17. Therefore, as previously shown (Chapter 7):

$$\sigma_c = \sqrt{17}$$
$$= 4.12$$

Since $k = 3$, and assuming a 95 percent confidence level, $h_p = 1.95$ (from Table 20-1). Therefore,

$$DL = \bar{c} \pm h_p\sigma_c$$
$$= 17 \pm 1.95(4.12)$$
$$= 17 \pm 8.0$$
$$= 9 \text{ to } 25$$

Consequently, formula A gives significantly poorer results, but formula C is the one of interest, yielding significantly better loom stops.

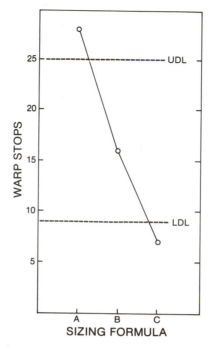

Fig. 20-2.   Analysis of warp stop rates using *DL*'s. Formula *C*, with the lowest stop rate, is significantly better.

An alternative approach might have been to record the loom stops per week for each week separately. The data might have then looked as follows:

	Sizing Formula			Grand Mean	Average Range
	A	*B*	*C*		
Week 1	6	4	2		
Week 2	4	2	3		
Week 3	9	8	1		
Week 4	9	2	1		
Mean	7	4	1.75	4.25	
Range	5	6	2		4.33

We may now look upon the rates as though they were "variables" data, and consider the weeks as replications, with $r = 4$. Furthermore,

Error Standard Deviation

$$\sigma = \overline{R}F_r$$
$$= 4.33(0.49)$$
$$= 2.12$$

based on the method of Chapter 19. To find DL's we note these further information items: (1) The total number N of observations is $k \times r = 3 \times 4 = 12$. (2) The $DF = k(r-1) = 3(4-1) = 9$. At a 95 percent level of confidence, Table 19-2 yields an h_d of 4.14:

$$DL = \overline{X} \pm h_d\sigma/\sqrt{N}$$
$$= 4.25 \pm 4.14(2.12)/\sqrt{12}$$
$$= 4.25 \pm 2.5$$
$$= 1.75 \text{ to } 6.75$$

Interestingly enough, these limits do not yield any significant differences. But the information gained from the experiment is still valuable. In particular, we note a relatively high range of variation for each sizing formula, suggesting several possible causes for detailed investigation: (1) uniformity of application of the size; (2) possible undue fluctuations in maintenance and operation of looms; (3) undue variation in the materials, such as the yarn quality and size constituents.

Note, however, that at a 90 percent level of confidence, h_d is 3.55. Hence

$$DL = 4.25 \pm 3.55(2.12)/\sqrt{12}$$
$$= 4.25 \pm 2.17$$
$$= 2.1 \text{ to } 6.4$$

indicating that size C is below the lower DL and thus significantly better than the average.

We conclude that (1) when significance cannot be established at one level of confidence, it may be established at a less demanding level; (2) even where an experiment yields no significant results, information benefits may nevertheless be gained. A somewhat controversial point is whether one should try to increase the sample size. It is not a pure process, statistically speaking, to do additional sampling when an experiment has not yielded significance and then add the sampling results to the experiment data already obtained. Theory says that an entirely new experiment, with larger N must now be run. In practice, however, most people interested in quick results will accept some statistical impurity and merely combine old with new data. The pressures of the real world to develop useful findings quickly cannot be ignored in practical experimentation, particularly when this is in-factory research. The need for sound, practical judgment remains.

Exponential Decision Lines

When studying differences in reliability among competing items for which exponential failure rates are expected to apply, interest is often confined to identifying the significantly superior item. This interest translates to a need for upper DL's only.

A study involved $k = 2$ competing designs, A and B. For each of these two items, $n = 50$ units were tested to observe the number of operating cycles to failure. A mean life m of 10,000 cycles to failure

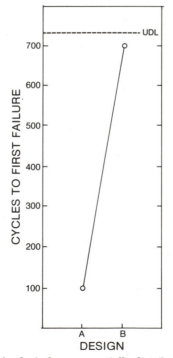

Fig. 20-3. Analysis for exponentially distributed failures.

had been specified, corresponding to a subgroup mean m' of $10,000/50 = 200$ cycles per test unit. Item A failed at 100 cycles, B failed at 700 cycles. At a 90 percent level of confidence, is B superior to A? We follow a simple series of steps:

1. Since there are $k = 2$ designs or "treatments" under comparison, find factor h_e corresponding to $k = 2$, from Table 20-2. In particular, $h_e = 2.99$.
2. $UDL = m' \times h_e$
   $= 200 \times 2.99$
   $= 598$ cycles
3. Design A at 700 cycles to failure is significantly better, by exceeding UDL.

As a further illustration, assume we had chosen a 95 percent confidence level. We would now find:

$$UDL = 200 \times 3.69$$
$$= 738 \text{ cycles}$$

so that 700 cycles is no longer superior at that level.

Inability to establish significance at a given level may mean that the sample size chosen was too small. Larger sample sizes increase

Table 20-2.   Factors h_e for Upper Decision Lines for the Exponential Distribution

Number k of Items Under Comparison	Confidence Level, Percent	
	90	95
2	2.99	3.69
3	3.41	4.10
4	3.69	4.39
5	3.91	4.60
6	4.10	4.79
7	4.25	4.94
8	4.39	5.08
9	4.51	5.19
10	4.60	5.30

SOURCE: Computed from the relationship
$e^{-UDL/m'} = 0.05/k$
for a 95 percent confidence level, with $1 - 0.95 = 0.05$.
For a 90 percent confidence level, $1 - 0.90 = 0.10$, replacing 0.05 in the formula above.
ACKNOWLEDGMENT: This application of Decision Lines for the Exponential Distribution is suggested by W. Schilling in "A Systematic Approach to the Analysis of Means, Part III," *J. of Quality Technology 5*, 4:156–8. (Oct. 1973).

the discriminating power of the *DL*'s. In the illustration just given, if n had been 100, rather than 50, we would find at the 95 percent confidence level:

$$m' = 10,000/100$$
$$= 100$$
$$UDL = 100 \times 3.69$$
$$= 369 \text{ cycles}$$

so that 700 cycles, being well above the *UDL* of 369, indicates a significantly superior reliability. Whether or not management can afford the cost of relatively large sample sizes depends on (1) the cost of reliability testing and (2) the economic importance of the particular research and the implications of its expected findings. These factors can vary considerably. As in all research, a moderate to high amount of possible failure may attend the investigational work.

Analysis Using Residuals

Consider the following results, in yield percent, for a production experiment in fermentation, using different amounts of aeration of the fermenter and two yeast strains, A and B[1]:

| | Aeration–Agitation | | | |
	Light	Medium	Heavy	Mean
Strain A	4.3	4.9	4.0	4.4
Strain B	4.5	4.1	4.0	4.2
Mean	4.4	4.5	4.0	4.3

Yield represents the proportion, in volume, of the alcohol distilled from the fermented mash of corn starch, water, and yeast.

One might try to use the method given for proportions of finding the error standard deviation, but this attempt will soon prove futile, since no sample size n in discrete units applies. Neither are the data of a Poisson or exponential type. They can, however, be treated as though they are variables. One caution applies: some people would

---

[1]Specialists in fermentation know that actual yields are 10 percent plus the percents shown above. However, removal of a constant does not affect the variances obtained; and for purposes of illustration we are using the values shown. Each value may be viewed, if one wishes as (10 + 4.X) percent, where X represents the decimals after 4. Our interest is centered on illustration of methodology, not on niceties of the state of the art of fermentation.

advise that percentages tend not to be normally distributed, and one should first convert the observations to polar coordinates. This is a complex approach, which we will omit, since the data, although not normal, are usually approximately normal.

There are no replications in this two-factor experiment. Hence we cannot estimate the error variance or the variation due to interaction effects. We can, however, obtain a so-called residual variation, in which the unknown magnitudes of error and interaction are contained as one combined figure.

Finding Residual Variation

Computer programs are available for calculating the residual variance and standard deviation, but a practical understanding of this term is best obtained by a direct demonstration. For this purpose, begin by subtracting the grand mean of 4.3 from the individual yield values. We find:

	Yield Values Expressed as Deviations from Grand Mean			New Mean
	0.0	0.6	−0.3	0.1
	0.2	−0.2	−0.3	−0.1
New Mean	0.1	0.2	−0.3	0.0

For example, first row, second entry, old value 4.9 subtracted from grand mean 4.3 yields a new value of 0.6. Unless there is an error in arithmetic or a rounding effect, the new grand mean will be zero. Now let us subtract the individual column means from each column entry. For the first column, $0.0 - 0.1 = -0.1$, $0.2 - 0.1 = 0.1$. Completing all three columns results in these values:

	Yield Values Expressed as Deviations from Column Means			New Mean
	−0.1	0.4	0.0	0.1
	0.1	−0.4	0.0	−0.1
New Mean	0.0	0.0	0.0	0.0

And, finally, let us subtract the row means from each row entry:

	Yield Values Expressed as Deviations from Row Means			New Mean
	−0.2	0.3	−0.1	0.0
	0.2	−0.3	0.1	0.0
New Mean	0.0	0.0	0.0	0.0

We are left with the residuals, which are now squared:

Squared Residuals		
-0.2^2	0.3^2	-0.1^2
-0.2^2	-0.3^2	0.1^2
0.04	0.09	0.01
0.04	0.09	0.01

equals

The sum of these squares is 0.28. The DF's based on this two-factor experiment, with three levels of one factor and two levels of the other factor, is $(3-1)(2-1) = 3$. Hence

$$\sigma^2_{\text{residuals}} = \Sigma(\text{residuals})^2/DF$$
$$= 0.28/3$$
$$= 0.093$$

and, taking the square root,

$$\sigma_{\text{residuals}} = 0.306$$

Given the absence of replication and the relatively small differences among the observed means, we can probably hope merely for significance at the 90 percent level of confidence. Now, $N = 6$, $DF = 3$. For $k = 2$ strains, h_d is 2.35, and for $k = 3$ aerations, $h_d = 5.29$. Now,

$$DL \text{ (strains)} = 4.3 \pm 2.35(0.306)/\sqrt{6}$$
$$= 4.3 \pm 0.96$$
$$= 3.34 \text{ to } 5.26$$

Therefore, the two strains are not significantly different. Finally,

$$DL \text{ (aerations)} = 4.3 \pm 5.29(0.306)/\sqrt{6}$$
$$= 4.3 \pm 0.66$$
$$= 3.64 \text{ to } 4.96$$

which is again not significant. Visual examination of the data, supported by Fig. 20-4, reveals the possibility of either strong interactions or error or both, since the lines do not run parallel. In such situations one cannot really expect to find significant results without replication.

Artificial Replication

Occasions arise when it is appropriate to create replication artificially, as a means of obtaining a useful error term. For example,

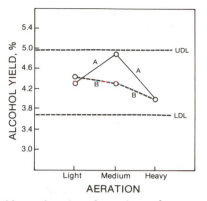

Fig. 20-4. Alcohol yield as a function of aeration and yeast strain. The nonparallel lines indicate (1) strong interaction, or (2) large experiment error, or (3) both. The main effects (aeration, strain) are not shown to be significant.

assume a situation in which advertising must occur within a limited space, and we can promote either product quality or price, but not both, while at the same time considering two types of audiences, rural and small town and city. The 400,000 promotions might then be applied as follows:

| Audience Type | Type of Product Emphasis | |
	Quality	Price
Rural	100,000	100,000
City	100,000	100,000

We can now code the circulars, such as by means of invisible ink, in binary form, such as 0 and 1. The mailing would then be:

| Audience Type | Type of Product Emphasis | |
	Quality	Price
Rural	50,000 coded zero	50,000 coded zero
	50,000 coded one	50,000 coded one
City	50,000 coded one	50,000 coded one
	50,000 coded zero	50,000 coded zero

We would then have a replication of two for each experiment cell, where the individual cell represents a particular product type and audience type combination.

Concluding Observations

The topic of experimentation in a search for improved product quality and reliability, for enhanced quality/cost ratio, or for achievement of higher productivity and reduced losses involves a variety of interesting approaches. We have presented some principal examples of application areas as a means of showing the potential in systematic research investigations. The manufacturing organization that wishes to stay competitive or desires to be a leader in technological innovation and growth will find it increasingly essential to create an environment in which research in the laboratory and in production becomes a routine aspect of operations, and in which experiment findings find effective translation into practical applications.

REVIEW QUESTIONS

1. Five competitive wood preservatives were investigated under simulated, accelerated conditions. For each preservative, 20 sample units of treated wood received laboratory aging for the equivalent of 10 years. The following showed a more than 50 percent loss in tensile strength:

	Preservative				
	A	*B*	*C*	*D*	*E*
Percent failing	5	8	2	10	25

Evaluate these results with the aid of 95 percent confidence *DL*'s.

2. Fabric woven by three different methods was inspected, and the number of defects per 1000 meters recorded:

	Conventional Loom	Air Jet Loom	Water Jet Loom
Defects per 1000 meters	20	12	31

Evaluate these performances in terms of 90 and 95 percent *DL*'s.

3. A knitted heart valve is expected to last the equivalent of five years' heartbeat. Two designs, one Teflon knitted and the other nylon knitted, are under study. Twenty valves of each type are tested. The first Teflon-knitted failure occurred at 2.25 years, and the first nylon-knitted failure occurred at 3.8 years. Evaluate the differences

at the 90 and 95 confidence levels, using DL's, and assuming an exponential failure distribution.

4. Find the residual variance for the following data relating to the reliability, in terms of probability of failure, percent for a rocket engine.

Inlet Pressure	Nozzle Design		
	Wide	Medium	Narrow
Low	5	5	8
High	3	1	2

Also, find DL's at the 90 percent confidence level and evaluate the observed failures against the DL's.

5. Can one run experiments without replication? Explain.
6. The following are data on the efficiency of dye-uptake in three types of dye-bath: 88, 75, and 80 percent. Why can these findings not be treated like a proportion for the purpose of calculating the standard deviation of the error?
7. For each of the following types of data indicate the probability distribution applicable:

   Breaking strength in kg/m.

   Mail order filling errors per week.

   Bee stings in Dade county in 1984.

   Number of insects per square yard of garden.

   Number of defective insulators in a shipment of 1000 insulators.

   Times to first failure of three lots, each consisting of 100 fluorescent lights.

8. What are some ways in which replication can be achieved?
9. Why can interaction not be determined separately when there are no replications?
10. What are the two components of variation comprising residual variation in the types of experiment presented in this chapter?

CHAPTER 21

Quality and Reliability Experiments Using Covariance Analysis

When decision line analysis fails in demonstrating a significant difference of individual treatment effects against the group average, a more advanced approach—covariance analysis—may be required. This method is more sensitive in discovering significant effects, but also involves greater complexities of application. Nevertheless, the essential principles are not difficult to grasp, and calculations can usually be carried out with suitable computer programs.

A Case History

One of the best illustrations of covariance analysis is due to B. B. Day, F. R. Del Priore, and E. Sax,[1] which we present here in adapted and slightly simplified form. A rust arrestor is applied by three different treatments: brushing (A), spraying (B), and dipping (C). The initial rust on each of five metal coupons is compared to the rust added during a salt spray. The data appear below:

	Treatment A: Brush		Treatment B: Spray	
	Added Rust Y_1	Original Rust X_1	Added Rust Y_2	Original Rust X_2
	29	11	52	38
	48	32	43	30
	50	40	24	13
	31	9	44	23
	32	8	47	31
Total:	190	100	210	135
Mean:	38	20	42	27

---

[1] "The Technique of Regression Analysis," *Quality Control Conference Papers* 1953, Milwaukee, Wisconsin: American Society for Quality Control.

	Treatment C: Dip		Cross Total	
	Added Rust Y_3	Original Rust X_3	Y_T	X_T
	40	30	121	79
	24	19	115	81
	27	29	101	82
	29	22	104	54
	20	10	99	49
Total:	140	110	540	345
Mean:	28	22	36	23

Five randomly chosen metal coupons were used for each of the three treatments, making a total of 15 coupons. The rust index values X before treatment and Y after treatment were determined.

Analysis Steps

Express each rust index as a deviation from the grand mean. The grand means for Y and X, respectively, are 36 and 29, so that the first entries for the deviations are $y = 29 - 36 = -7$ and $x = 11 - 23 = -12$. For the entire set of 15 pairs, the tabulation below results:

Data Shown as Deviations From Grand Means (36, 23)

	y_1	x_1	y_2	x_2	y_3	x_3	y_t	x_t
	-7	-12	16	15	4	7	13	10
	$+12$	9	7	7	-12	-4	7	12
	$+14$	17	-12	-10	-9	6	-7	13
	-5	-14	8	0	-7	-1	-4	15
	-4	-15	11	8	-16	-13	-9	-20
Total:	10	-15	30	20	-40	-5	0	0
Mean:	2	-3	6	4	-8	-1		

It is also possible to express each Y and X in terms of deviations from the respective column mean. For the Y_1 column, with mean 38, the first entry becomes $y = 29 - 38 = -9$. For the X_1 column, with column mean 20, the first entry is $11 - 20 = -9$. If the calculations are correct, each of the column sums must be zero. Again tabulate the results:

Data Shown as Deviations from Column Means

y_1	x_1	y_2	x_2	y_3	x_3
-9	-9	10	11	12	8
10	12	1	3	-4	-3
12	20	-18	-14	-1	7
-7	-11	2	-4	1	0
-6	-12	5	4	-8	12
Total: 0	0	0	0	0	0

Now find the squares and cross products for the data just tabulated.

Squares and Cross Products for Deviations
from Column Means

y_1^2	x_1^2	y_1x_1	y_2^2	x_2^2	y_2x_2	y_3^2	x_3^2	y_3x_3
81	81	81	100	121	110	144	64	96
100	144	120	1	9	3	16	9	12
144	400	240	324	196	252	1	49	-7
49	121	77	4	16	-8	1	0	0
36	144	72	25	16	20	64	144	96
Total 410	890	590	454	358	377	226	266	197

Totals

y^2	x^2	xy
1090	1514	1164

For the first entry for example, $y^2 = (-9)^2 = 81$. For the first cross total, $yx = (-9) \times (-9) = 81$.

The totals and means from the calculations involving x's and y's are now entered in a master analysis table, as shown on p. 317. The step-by-step procedures are self-explanatory. Only a few details need further exposition:

1. Number of observations: There are $k = 3$ treatments and $n = 5$ replications per treatment, so that $N = kn = 3 \times 5 = 15$.
2. Degrees of Freedom, DF: DF allows for the fact that, in calculating variances, sample means are used instead of the (unknown) population means. For differences among the $k = 3$ treatments, $DF = k - 1 = 3 - 1 = 2$. For experimental error, $DF = N - k - 1 = 15 - 3 - 1 = 11$. For the total column, DF is the sum of the two DF's just found, that is $2 + 11 =$

Table 21-1a. Critical Values of the *F*-Ratio for Significance Testing (95 Percent Confidence Level)

Degrees of Freedom for Numerator

Degrees of Freedom for Denominator	1	2	3	4	5	6	7	8	9	10	12	15	20	24	30	40	60	120	∞
1	161.4	199.5	215.7	224.6	230.2	234.0	236.8	238.9	240.5	241.9	243.9	245.9	248.0	249.1	250.1	251.1	252.2	253.3	254.3
2	18.51	19.00	19.16	19.25	19.30	19.33	19.35	19.37	19.38	19.40	19.41	19.43	19.45	19.45	19.46	19.47	19.48	19.49	19.50
3	10.13	9.55	9.28	9.12	9.01	8.94	8.89	8.85	8.81	8.79	8.74	8.70	8.66	8.64	8.62	8.59	8.57	8.55	8.53
4	7.71	6.94	6.59	6.39	6.26	6.16	6.09	6.04	6.00	5.96	5.91	5.86	5.80	5.77	5.75	5.72	5.69	5.66	5.63
5	6.61	5.79	5.41	5.19	5.05	4.95	4.88	4.82	4.77	4.74	4.68	4.62	4.56	4.53	4.50	4.46	4.43	4.40	4.36
6	5.99	5.14	4.76	4.53	4.39	4.28	4.21	4.15	4.10	4.06	4.00	3.94	3.87	3.84	3.81	3.77	3.74	3.70	3.67
7	5.59	4.74	4.35	4.12	3.97	3.87	3.79	3.73	3.68	3.64	3.57	3.51	3.44	3.41	3.38	3.34	3.30	3.27	3.23
8	5.32	4.46	4.07	3.84	3.69	3.58	3.50	3.44	3.39	3.35	3.28	3.22	3.15	3.12	3.08	3.04	3.01	2.97	2.93
9	5.12	4.26	3.86	3.63	3.48	3.37	3.29	3.23	3.18	3.14	3.07	3.01	2.94	2.90	2.86	2.83	2.79	2.75	2.71
10	4.96	4.10	3.71	3.48	3.33	3.22	3.14	3.07	3.02	2.98	2.91	2.85	2.77	2.74	2.70	2.66	2.62	2.58	2.54
11	4.84	3.98	3.59	3.36	3.20	3.09	3.01	2.95	2.90	2.85	2.79	2.72	2.65	2.61	2.57	2.53	2.49	2.45	2.40
12	4.75	3.89	3.49	3.26	3.11	3.00	2.91	2.85	2.80	2.75	2.69	2.62	2.54	2.51	2.47	2.43	2.38	2.34	2.30
13	4.67	3.81	3.41	3.18	3.03	2.92	2.83	2.77	2.71	2.67	2.60	2.53	2.46	2.42	2.38	2.34	2.30	2.25	2.21
14	4.60	3.74	3.34	3.11	2.96	2.85	2.76	2.70	2.65	2.60	2.53	2.46	2.39	2.35	2.31	2.27	2.22	2.18	2.13

15	4.54	3.68	3.29	3.06	2.90	2.79	2.71	2.64	2.59	2.54	2.48	2.40	2.33	2.29	2.25	2.20	2.16	2.11	2.07
16	4.49	3.63	3.24	3.01	2.85	2.74	2.66	2.59	2.54	2.49	2.42	2.35	2.28	2.24	2.19	2.15	2.11	2.06	2.01
17	4.45	3.59	3.20	2.96	2.81	2.70	2.61	2.55	2.49	2.45	2.38	2.31	2.23	2.19	2.15	2.10	2.06	2.01	1.96
18	4.41	3.55	3.16	2.93	2.77	2.66	2.58	2.51	2.46	2.41	2.34	2.27	2.19	2.15	2.11	2.06	2.02	1.97	1.92
19	4.38	3.52	3.13	2.90	2.74	2.63	2.54	2.48	2.42	2.38	2.31	2.23	2.16	2.11	2.07	2.03	1.98	1.93	1.88
20	4.35	3.49	3.10	2.87	2.71	2.60	2.51	2.45	2.39	2.35	2.28	2.20	2.12	2.08	2.04	1.99	1.95	1.90	1.84
21	4.32	3.47	3.07	2.84	2.68	2.57	2.49	2.42	2.37	2.32	2.25	2.18	2.10	2.05	2.01	1.96	1.92	1.87	1.81
22	4.30	3.44	3.05	2.82	2.66	2.55	2.46	2.40	2.34	2.30	2.23	2.15	2.07	2.03	1.98	1.94	1.89	1.84	1.78
23	4.28	3.42	3.03	2.80	2.64	2.53	2.44	2.37	2.32	2.27	2.20	2.13	2.05	2.01	1.96	1.91	1.86	1.81	1.76
24	4.26	3.40	3.01	2.78	2.62	2.51	2.42	2.36	2.30	2.25	2.18	2.11	2.03	1.98	1.94	1.89	1.84	1.79	1.73
25	4.24	3.39	2.99	2.76	2.60	2.49	2.40	2.34	2.28	2.24	2.16	2.09	2.01	1.96	1.92	1.87	1.82	1.77	1.71
26	4.23	3.37	2.98	2.74	2.59	2.47	2.39	2.32	2.27	2.22	2.15	2.07	1.99	1.95	1.90	1.85	1.80	1.75	1.69
27	4.21	3.35	2.96	2.73	2.57	2.46	2.37	2.31	2.25	2.20	2.13	2.06	1.97	1.93	1.88	1.84	1.79	1.73	1.67
28	4.20	3.34	2.95	2.71	2.56	2.45	2.36	2.29	2.24	2.19	2.12	2.04	1.96	1.91	1.87	1.82	1.77	1.71	1.65
29	4.18	3.33	2.93	2.70	2.55	2.43	2.35	2.28	2.22	2.18	2.10	2.03	1.94	1.90	1.85	1.81	1.75	1.70	1.64
30	4.17	3.32	2.92	2.69	2.53	2.42	2.33	2.27	2.21	2.16	2.09	2.01	1.93	1.89	1.84	1.79	1.74	1.68	1.62
40	4.08	3.23	2.84	2.61	2.45	2.34	2.25	2.18	2.12	2.08	2.00	1.92	1.84	1.79	1.74	1.69	1.64	1.58	1.51
60	4.00	3.15	2.76	2.53	2.37	2.25	2.17	2.10	2.04	1.99	1.92	1.84	1.75	1.70	1.65	1.59	1.53	1.47	1.39
120	3.92	3.07	2.68	2.45	2.29	2.17	2.09	2.02	1.96	1.91	1.83	1.75	1.66	1.61	1.55	1.50	1.43	1.35	1.25
∞	3.84	3.00	2.60	2.37	2.21	2.10	2.01	1.94	1.88	1.83	1.75	1.67	1.57	1.52	1.46	1.39	1.32	1.22	1.00

NOTES: 1. To be significant, an observed F must equal or exceed the tabulated critical F.

2. Example: $DF_{numerator} = 2$, $DF_{denominator} = 11$, $F_{observed} = 11.2$.

Then, since the tabular F is 3.98, we concluded that the observed F is significant at 95 percent confidence. It might also be significant at 99 percent confidence (compare against the critical F of Table 21-1b).

Table 21-1b. Critical Values of the F-Ratio for Significance Testing
(99 Percent Confidence Level)

Degrees of Freedom for Denominator

Degrees of Freedom for Denominator	1	2	3	4	5	6	7	8	9	10	12	15	20	24	30	40	60	120	∞
1	4052	4999.5	5403	5625	5764	5859	5928	5982	6022	6056	6106	6157	6209	6235	6261	6287	6313	6339	6366
2	98.50	99.00	99.17	99.25	99.30	99.33	99.36	99.37	99.39	99.40	99.42	99.43	99.45	99.46	99.47	99.47	99.48	99.49	99.50
3	34.12	30.82	29.46	28.71	28.24	27.91	27.67	27.49	27.35	27.23	27.05	26.87	26.69	26.60	26.50	26.41	26.32	26.22	26.13
4	21.20	18.00	16.69	15.98	15.52	15.21	14.98	14.80	14.66	14.55	14.37	14.20	14.02	13.93	13.84	13.75	13.65	13.56	13.46
5	16.26	13.27	12.06	11.39	10.97	10.67	10.46	10.29	10.16	10.05	9.89	9.72	9.55	9.47	9.38	9.29	9.20	9.11	9.02
6	13.75	10.92	9.78	9.15	8.75	8.47	8.26	8.10	7.98	7.87	7.72	7.56	7.40	7.31	7.23	7.14	7.06	6.97	6.88
7	12.25	9.55	8.45	7.85	7.46	7.19	6.99	6.84	6.72	6.62	6.47	6.31	6.16	6.07	5.99	5.91	5.82	5.74	5.65
8	11.26	8.65	7.59	7.01	6.63	6.37	6.18	6.03	5.91	5.81	5.67	5.52	5.36	5.28	5.20	5.12	5.03	4.95	4.86
9	10.56	8.02	6.99	6.42	6.06	5.80	5.61	5.47	5.35	5.26	5.11	4.96	4.81	4.73	4.65	4.57	4.48	4.40	4.31
10	10.04	7.56	6.55	5.99	5.64	5.39	5.20	5.06	4.94	4.85	4.71	4.56	4.41	4.33	4.25	4.17	4.08	4.00	3.91
11	9.65	7.21	6.22	5.67	5.32	5.07	4.89	4.74	4.63	4.54	4.40	4.25	4.10	4.02	3.94	3.86	3.78	3.69	3.60
12	9.33	6.93	5.95	5.41	5.06	4.82	4.64	4.50	4.39	4.30	4.16	4.01	3.86	3.78	3.70	3.62	3.54	3.45	3.36
13	9.07	6.70	5.74	5.21	4.86	4.62	4.44	4.30	4.19	4.10	3.96	3.82	3.66	3.59	3.51	3.43	3.34	3.25	3.17

14	8.86	6.51	5.56	5.04	4.69	4.46	4.28	4.14	4.03	3.94	3.80	3.66	3.51	3.43	3.35	3.27	3.18	3.09	3.00
15	8.68	6.36	5.42	4.89	4.56	4.32	4.14	4.00	3.89	3.80	3.67	3.52	3.37	3.29	3.21	3.13	3.05	2.96	2.87
16	8.53	6.23	5.29	4.77	4.44	4.20	4.03	3.89	3.78	3.69	3.55	3.41	3.26	3.18	3.10	3.02	2.93	2.84	2.75
17	8.40	6.11	5.18	4.67	4.34	4.10	3.93	3.79	3.68	3.59	3.46	3.31	3.16	3.08	3.00	2.92	2.83	2.75	2.65
18	8.29	6.01	5.09	4.58	4.25	4.01	3.84	3.71	3.60	3.51	3.37	3.23	3.08	3.00	2.92	2.84	2.75	2.66	2.57
19	8.18	5.93	5.01	4.50	4.17	3.94	3.77	3.63	3.52	3.43	3.30	3.15	3.00	2.92	2.84	2.76	2.67	2.58	2.49
20	8.10	5.85	4.94	4.43	4.10	3.87	3.70	3.56	3.46	3.37	3.23	3.09	2.94	2.86	2.78	2.69	2.61	2.52	2.42
21	8.02	5.78	4.87	4.37	4.04	3.81	3.64	3.51	3.40	3.31	3.17	3.03	2.88	2.80	2.72	2.64	2.55	2.46	2.36
22	7.95	5.72	4.82	4.31	3.99	3.76	3.59	3.45	3.35	3.26	3.12	2.98	2.83	2.75	2.67	2.58	2.50	2.40	2.31
23	7.88	5.66	4.76	4.26	3.94	3.71	3.54	3.41	3.30	3.21	3.07	2.93	2.78	2.70	2.62	2.54	2.45	2.35	2.26
24	7.82	5.61	4.72	4.22	3.90	3.67	3.50	3.36	3.26	3.17	3.03	2.89	2.74	2.66	2.58	2.49	2.40	2.31	2.21
25	7.77	5.57	4.68	4.18	3.85	3.63	3.46	3.32	3.22	3.13	2.99	2.85	2.70	2.62	2.54	2.45	2.36	2.27	2.17
26	7.72	5.53	4.64	4.14	3.82	3.59	3.42	3.29	3.18	3.09	2.96	2.81	2.66	2.58	2.50	2.42	2.33	2.23	2.13
27	7.68	5.49	4.60	4.11	3.78	3.56	3.39	3.26	3.15	3.06	2.93	2.78	2.63	2.55	2.47	2.38	2.29	2.20	2.10
28	7.64	5.45	4.57	4.07	3.75	3.53	3.36	3.23	3.12	3.03	2.90	2.75	2.60	2.52	2.44	2.35	2.26	2.17	2.06
29	7.60	5.42	4.54	4.04	3.73	3.50	3.33	3.20	3.09	3.00	2.87	2.73	2.57	2.49	2.41	2.33	2.23	2.14	2.03
30	7.56	5.39	4.51	4.02	3.70	3.47	3.30	3.17	3.07	2.98	2.84	2.70	2.55	2.47	2.39	2.30	2.21	2.11	2.01
40	7.31	5.18	4.31	3.83	3.51	3.29	3.12	2.99	2.89	2.80	2.66	2.52	2.37	2.29	2.20	2.11	2.02	1.92	1.80
60	7.08	4.98	4.13	3.65	3.34	3.12	2.95	2.82	2.72	2.63	2.50	2.35	2.20	2.12	2.03	1.94	1.84	1.73	1.60
120	6.85	4.79	3.95	3.48	3.17	2.96	2.79	2.66	2.56	2.47	2.34	2.19	2.03	1.95	1.86	1.76	1.66	1.53	1.38
∞	6.63	4.61	3.78	3.32	3.02	2.80	2.64	2.51	2.41	2.32	2.18	2.04	1.88	1.79	1.70	1.59	1.47	1.32	1.00

SOURCE: Adapted from M. Merrington and C. M. Thompson, "Tables of Percentage Points of the Inverted Beta (F) Distribution," *Biometrika* **38** (2), 73–88 (1943). Abbreviated, only two instead of four decimals used.

13. As a check, total DF is found directly from $N - 2 = 15 - 2 = 13$.

3. Sums of squares. For y_1 the individual means of 2, 6, and -8 square to 4, 36, and 64, which totals 104. Next $n(104) = 5(104) = 520$. Proceed similarly for x and xy. For the error column, find the totals for y^2, x^2, and xy from the tabulation for squares and cross products. The total column is the sum of the two individual columns. For row c for example, $520 + 1090 = 1610$.

4. Regression coefficient. This step is self-explanatory.

5. Adjusted sums of squares (row h): self-explanatory.

6. Mean square: This is a type of variance, obtained by the self-explanatory steps demonstrated.

7. F-ratio. Obtained by dividing the mean square for the differences among treatments by the error mean square. The larger this ratio, the more likely is it that the effect of treatments is significant.

8. Critical F: At the 99 percent confidence level, Table 21-1b shows 7.21 as the critical F. The observed F must equal or exceed the critical F in order to establish significance of difference among the treatments.

Since the observed $F = 11.2$ exceeds F-critical of 7.21, we conclude at the 99 percent level of confidence that the differences among the three rust arrestor applications are significant, with treatment C: dip, best, since it adds the least amount of rust after salt spray.

Graphic Covariance

For a full understanding of the results of an experiment, it is almost always essential to graph the important data. This is done in Fig. 21-1 for the covariance analysis. While the coordinates could have been given in terms of X and Y, we found it preferable to show instead x and y. The plotted points form what is known as a "scatter diagram."

The average relationship between x and y is shown by the so-called regression line, developed separately for treatments A, B, and C. Again treatment C, with the regression line running well below the other two, is demonstrated as the best rust arrestor.

Comparison with Decision Line Analysis

How would DL's compare with covariance analysis? For the purpose of decision lines, we would examine only the index of additional

Covariance Analysis Master Table

Procedure and Steps	Sources of Variation		
	Differences Among Columns, C	Experimental Error, E	Total, T
a. Number of Observations	$k = 3$ Treatments	$n = 5$ trials/treatment	$N = 3 \times 5 = 15$
b. Degrees of Freedom, DF	$DF = 3-1 = 2$	$DF = N-k-1 = 15-3-1 = 11$	$DF = 15-2 = 13$
c. n(Sum of $\bar{y}$-squared)	$5(2^2 + 6^2 - 8^2) = 520$	1090^*	1610^{**}
d. n(Sum of $\bar{x}$-squared)	$5(-3^2 + 4^2 + -1^2) = 130$	1514	1644
e. n(Sum of $\bar{x}\,\bar{y}$)	$5[2(-3) + 6(4) + -8(-1)] = 130$	1164	1294
f. Regression Coefficient $= e/d$		0.7688	0.7871
g. Amount of Variation Ascribable to and Explained by Regression $= e^2/d$		894.9	1018.5
h. Sums of Squares for y, adjusted for effect of Regression	$c_C - (g_T - g_E) = 520 - (1018.5 - 894.9)$ $= 396.4$	$c - g$ $= 195.1$	$c - g$ $= 591.5$
i. Mean-square, h/b	198.2	17.7	
j. F-ratio $= i_C/i_E$	$198.2/17.7 = 11.2$		
k. Critical F at 99% Confidence Level (From Table 21-1b)	7.21		—

*The three entries in this column are $\Sigma y^2 = 1090$, $\Sigma x^2 = 1514$, and $\Sigma xy = 1164$.

**Unless otherwise indicated the totals are from columns C and E.

rust, using the familiar method of analysis shown below:

Decision Line Analysis of Rust Arrestor Treatment
(Index of Additional Rust After Salt Exposure of 15 Metal Coupons)

	Treatment A: Brush-on of Arrestor	Treatment B: Spray-on of Arrestor	Treatment C: Dip-Use of Arrestor	Other
	29	52	40	
	48	43	24	
	50	24	27	
	31	44	29	
	32	47	20	
Mean	38	42	22	
Standard Deviation	14.9	9.5	8.2	
Average of Standard Deviations = 32.6/3				10.87
Number of treatments, k				3
Degrees of Freedom = 3(5−1)				12
Factor h_d at 95% Confidence				3.93
Grand Mean = 102/3				34
Decision Line = 34 ± 3.93(10.87)/$\sqrt{15}$				34 ± 11

Only treatment C at a rust index of 22 falls outside the DL's of 23–45, and this at the 95 percent confidence level. On the other hand, covariance yielded an observed F that was well above the 99 percent confidence level.

General Covariance Lesson

A general lesson applies from our illustrative example: where applicable, covariance is a more sensitive test of significance than other methods, such as decision line analysis. In practice this often means that when ordinary methods do not reveal significance among averages, a covariance analysis may yet be successful.

The drawback of covariance analysis is that (1) data for performing this analysis may not be available (such as the Y values corresponding to the X's) and (2) that it is more complex and thus more costly to apply. The computational aspects of this more complex task may be minimized with use of an appropriate computer program. Some might argue that covariance analysis is also more difficult to un-

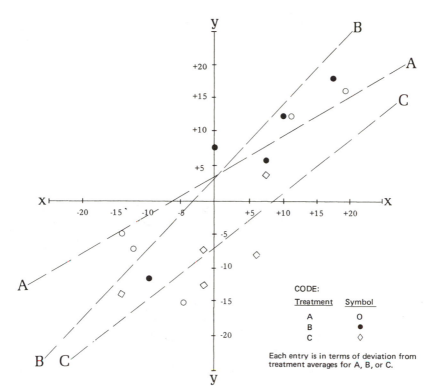

Fig. 21-1.   Rust arrestor effect: graphic comparison among three treatments: *A*, *B*, and *C*.

derstand. But this aspect applies only to the design and computational requirements of the method. The analysis results, when presented in terms of (1) confidence level associated with significance and (2) graphic portrayal of the relationships investigated, are quite simple to comprehend.

It is of course axiomatic to all data analyses that, when one approach fails to establish significant results, alternative approaches may be more applicable. When the analysis and evaluation problem becomes unduly complex and cumbersome, it is wise to seek aid from specialists in the design and analysis of experiments.

The Variance Analysis Alternative

Having examined the *DL* method as an alternative to covariance analysis, let us now consider analysis of variance, based on the individual y's as deviations from the grand mean 36 for all Y's:

y_1	y_2	y_3	Grand Mean
−7	16	4	
12	7	−12	
14	−12	−9	
−5	8	−7	
−4	11	−16	
Mean: 2	6	−8	0

Let us square these values:

y_1^2	y_2^2	y_3^2	
49	256	16	
144	49	144	
196	144	81	
25	64	49	
16	121	256	
Total 430	634	546	1610

This total Sum of Squares of 1610 also appears in line c of the Covariance Analysis Master Table, last column. That same line, under the column E, shows the Experimental Error Sum of Squares of 1090. Finally, under C for line c is the Between Treatments Sum of Squares of 520. The DF's remain 2 for treatments, 11 for experimental error, and 13 for total. Putting these results into a summary table, we find:

Analysis of Variance for Rust Arrestor Experiment

(a) Source of Variation	(b) Sum of Squares	(c) DF	(d) Mean Square = b/c	(e) F-ratio
Total	1610	13	—	
Treatments	520	2	260	
Error	1090	11	99	
Observed F = 260/99				2.62
Critical F (at 95 per-cent Confidence Level, from Table 21-1a)				3.98

Therefore, the observed F is less than the critical F, and we conclude that the differences are not significant.

At this point some may wonder: How is it that the DL's showed statistical significance at 95 percent confidence, but the analysis of variance fails in showing significance? The answer is that the two methods ask slightly different questions. *DL* analysis asks: Is any individual average significantly higher or lower than the grand mean? Analysis of variance asks: Taken in totality, do the means differ among each other? In the first question, statistical evaluation answered "yes," and in the second question it answered "no."

This finding underscores our previous observation: If one statistical method fails in demonstrating significance, another relevant method may yet succeed.

Calculation of Regression Line

In order to show how a regression line is calculated, let us examine the relationship between x and y for treatment C. From the tabulation of squares and cross products, we find: Sum of $x_3^2 = 64 + 9 + 49 + 0 + 144 = 266$. Sum of $y_3x_3 = 96 + 12 + 7 + 0 + 76 = 191$. Therefore, the coefficient b of the regression line

$$y' = a + bx$$

is found from the ratio

$$b = \Sigma xy/\Sigma x^2$$
$$= 191/266$$
$$= 0.718$$

The regression coefficient b is the slope of the regression line $y' = a + bx$. For treatment C, as an example, a is found from

$$a = y' - b\bar{x}$$
$$= -8 - 0.718(-1)$$
$$= -7.288$$

which we enter at the point $x = 0$ on the graph. We need one more point of y to draw a line. Assume $x = 20$, then

$$y' = a + bx$$
$$= -7.288 + 0.718(20)$$
$$= 7.132$$

Connecting -7.288 at $x = 0$ with 7.132 at $x = 20$ yields the regression line.

As an incidental observation, it now becomes clear why the total degrees of freedom in the covariance analysis was $N - 2$. Since any two points form a line, the degrees of freedom is the total number of points N less this minimum of two points, hence $N - 2$.

Calibration Application

Regression analysis has many valuable uses. It is thus appropriate to present a further example, taken from instrument calibration. A paper manufacturer was testing the moisture content of incoming pulp by the conventional method of a drying oven and an electronic moisture meter. How well did the moisture meter readings calibrate with the more reliable, but also more costly and time consuming, drying oven?

The six observation pairs of moisture gage X and drying oven Y results will represent, for illustrative purposes, the 20–30 pairs normally used in a calibration or other regression study. The calculations are self-explanatory from the tabulation below.

Calibration Study: Original Data, Deviations, Squares, and Covariance

Lot Number	Meter, X	Oven, Y	Deviations from Means		Squares		Covariance, xy
			x	y	x^2	y^2	
1	7.9	6.6	1.9	1.6	3.61	2.56	3.04
2	5.3	3.4	−0.7	−1.6	0.49	2.56	1.12
3	5.8	4.2	−0.2	−0.8	0.04	0.64	0.16
4	6.0	4.8	0	−0.2	0	0.04	0
5	4.8	5.0		0	1.44	0	0
6	6.2	6.0		1.0	0.04	1.00	0.20
Totals	36.0	30.0	0	0	5.62	6.80	4.52

$b = 4.52/5.62 = 0.80$
$a = \bar{Y} - b\bar{X} = 5 - 0.8(6) = 0.2$
Y' (when $X = 4$) $= 3.4$; Y' (when $X = 8$) $= 6.6$

The regression line is inserted into the so-called scatter plot of the observation pairs, in Fig. 21-2. The fact that the regression line falls below a theoretical line of equality, in which gage and oven agree fully, indicates a bias toward high readings on the part of the gage. The average overstatement of moisture ranges from 1 to 1.4 percent. This bias represents lack of accuracy. We will next examine precision.

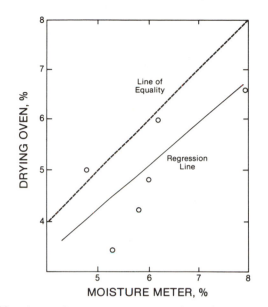

Fig. 21-2.   Calibration, moisture content, percent by moisture meter versus drying oven. The regression line falls below the theoretical equality line (moisture meter = drying oven). The difference represents bias or lack of accuracy of the moisture meter (the meter tends to read higher than the oven).

Precision

Precision in regression analysis is represented by the standard error of estimate:

$$\text{Standard Error of Estimate} = (\Sigma x^2 - b\Sigma xy)^{1/2}$$
$$= \sqrt{5.62 - 0.8(4.52)}$$
$$= 1.4$$

Now, since ± 2 standard deviations or ± 2 standard errors covers 95 percent of a normal distribution, we can say

$$\text{Precision (at 95 percent confidence)} = \pm\,2 \times 1.4$$
$$= \pm\,2.8$$

in terms of moisture, percent. Actually, in this case, this precision represents a relatively wide error. Combined with the poor accuracy, management is probably best advised to seek a moisture gage working on some alternative principle.

Correlation

A question closely allied to precision is "How good is the correlation; that is, how well do the individual points fit around the regression line?" We find the correlation coefficient r from

$$r^2 = \Sigma xy/(\Sigma x^2 \Sigma y^2)^{1/2}$$
$$= 4.52/\sqrt{5.62 \times 6.8}$$
$$r^2 = 0.73$$
$$r = 0.86$$

The degrees of freedom for these $n = 6$ observation pairs is $n - 2, = 6 - 2 = 4$. From Table 21-2 of Minimum Values of the Correlation Coefficient Needed to Establish Statistical Significance, we observe the following critical values:

	Confidence Level	
	90%	95%
Correlation coefficient	0.73	0.81

Table 21-2.   Minimum Values of the Correlation Coefficient r Needed to Establish Statistical Significance

Degrees of Freedom	Confidence Level (%)		
	90	95	99
1	0.99	1.00	1.00
2	0.90	0.95	0.99
3	0.81	0.88	0.96
4	0.73	0.81	0.92
5	0.67	0.75	0.87
6	0.62	0.71	0.83
8	0.55	0.63	0.77
10	0.50	0.58	0.71
16	0.41	0.48	0.61
20	0.36	0.42	0.54
25	0.32	0.38	0.49
30	0.30	0.35	0.50
40	0.26	0.30	0.39
50	0.23	0.27	0.35
60	0.21	0.25	0.33
80	0.18	0.22	0.28
100	0.16	0.20	0.25

NOTES: Degrees of Freedom DF in simple correlation refers to 2 less than the number n of plotted pairs of points. Hence, $DF = n - 2$.
SOURCE: Calculated on the basis of a statistical relationship between r^2 and F, and referring to tables of the F-ratio.

The observed value $r = 0.86$, thus establishing significance. Thus there is correlation, even though in our example the relatively high standard error of estimate makes the regression line only approximately useful.

Regression and Correlation, General Observations

The topic of regression and correlation is vast. In practical reliability and quality control work there are many applications. A few examples are:

1. In developing a ballistic propellant, what is the effect of various proportions of nitrocellulose on impact modulus?
2. What is the relationship between weld diameter and shear strength for a particular type of weld?
3. An accelerated test has been developed. What is the relationship between test-predicted life of a device and actual field experience of satisfactory operation until failure?

Relationships may involve several variables. For example:

1. In evaluating the field quality of armor plate, what is the effect of thickness and Brinell hardness of the metal on the plate's ballistic limit?
2. How are fiber length, strength, and fineness related to the resultant yarn strength?
3. What is the effectiveness of a detergent under varying time cycles, concentrations, and washing temperatures?
4. How do formulation, heating cycle, and thickness of application of a glossy coating affect the performance life of an electrical insulator?

These types of analysis do not affect the day-to-day operation of quality control and reliability assurance, but they are a part of the underlying work that must occur in product design and process development to be able to market a superior product.

Concluding Observations

We have presented covariance analysis and the related topic of regression and correlation. The statistical methods involved are more complex than those for simple control charting or the application of sampling plans. They are useful in those situations where it is desired to evaluate competitive raw materials, product designs, or processing methods, so as to come up with a decision that is optimal from a viewpoint of cost, quality, and reliability in the overall manufacturing cycle, from conception to consumer. While the quality control engineer or technologist may not be called on directly to perform these types of analyses, he or she should nevertheless be aware of their essential nature and applicability. This will permit such a person (1) to identify cases in which covariance or regression analysis may be helpful in providing insights for product or process improvement and (2) to interact effectively with statistical specialists in the evaluation of analysis results.

REVIEW QUESTIONS

1. What is the difference between decision line analysis and covariance analysis? Compare advantages and disadvantages.
2. What is the difference between a line of equality and a regression line?
3. State an alternative name for "regression line."
4. What is the purpose of the F-test?
5. The following data show adhesive thickness in 0.01 cm and cohesion strength in kg for two products A and B:

	Adhesive A		Adhesive B	
	Thickness	Strength	Thickness	Strength
	10	22	11	22
	12	26	9	20
	11	21	10	21
Mean	11	23	10	21

For the six specimens tested, is there a significant difference among strengths between A and B? In answering this question, perform a DL analysis and a covariance analysis.

4. For eight lots, a comparison of laboratory tested cycles to failure (in 1000's) and subsequent field experience was as follows:

Lot No.	1	2	3	4	5	6	7	8
Laboratory Sample:	25	19	20	22	28	18	20	24
Field Experience:	21	16	18	19	25	15	18	20

Evaluate the precision and accuracy of the laboratory method with the aid of regression analysis.

5. A pharmaceutical firm, working with a new antibiotic, is well aware that the inhibition circle of antibiotic, applied to a germ culture,

increases in curvilinear relation to the concentration of the antibiotic; based on these observations:

Concentration, units/ml, Z	Alternate Form of Z	Exponent of 2.0, X	Inhibition Circle, Y
1	2^0	0	4
2	2^1	1	5
4	2^2	2	9
8	2^3	3	10
16	2^4	4	12

Correlate X and Y. Using Z as a base, plot Y. Next, using X as a base, plot Y.

CHAPTER 22

Special Methods for Participative Quality-Control Activities

The realization that important gains in quality, productivity, and cost reduction can accrue from motivated worker participation in the decision process is American in origin. The first controlled study in this regard is the well-known Western Electric Hawthorne Works series of 1927–1932. It remained for Japanese industry—through the efforts of the Japanese Union of Scientists and Engineers, the Japan Employers' Association, and the personal efforts of Dr. Kaoru Ishikawa, Professor of Engineering at the University of Tokyo—to drive home to the world the full value of a good participative management program.

Quality Circles

The motivative–participative approach has found its greatest use in the form of Quality Circles. In each sector of a manufacturing or service organization, a group of workers and their supervisor organize themselves in a "circle" for the purpose of reviewing problems related to quality, cost, productivity, and safety, with a view to seek jointly for an effective solution. Simple graphic approaches, such as control charts, frequency pattern diagrams, and *DL*'s are particularly suited for quality circles, since the problem information is presented in clear, readily comprehensive form in well-prepared graphs.

A number of additional specialized techniques—some old, some new—have been of further importance in the success of the work of quality circles, as discussed below.

Pareto Diagrams

Alfred Pareto (1848–1923), an economist, developed a concept known as Pareto's Law. In simplified form it states that 80 percent of all

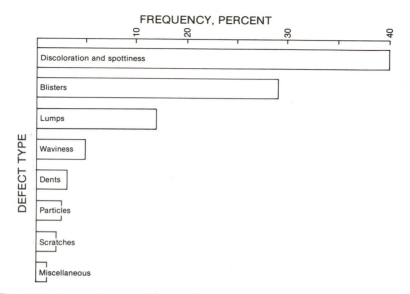

Fig. 22-1. Pareto diagram. The few dominant and the many trivial (infrequent) defectives.

problems stem from 20 percent of the causes. For example,

80 percent of the inventory value of a firm is in 20 percent of the inventory items.

80 percent of all the rejections, returns, rework, and allowances come from 20 percent of the items made.

80 percent of sales volume comes from 20 percent of the customers.

The Law must not be taken too literally, and certainly the percentages shown will vary. They are true only in a very general sense.

An illustrative Pareto Diagram appears in Fig. 22-1. For a particular product, most of the quality problems (discoloration, blisters) come from only two out of eight or more possible areas. A group examining a Pareto graph will thus recognize quickly which types of problems predominate, that these are only a few, and that taking care of these few takes care of 80 percent of all causes of the problem situation.

Cause-and-Effect Diagram

A technique for systematically listing the various causes of a problem, credited to Dr. Kaoru Ishikawa, is the Cause-and-Effect Dia-

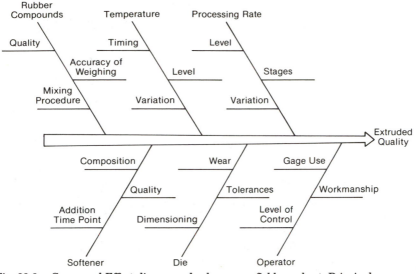

Fig. 22-2. Cause-and-Effect diagram, also known as *fishbone* chart. Principal sources of potential quality problems are drawn in under applicable categories. With experience, additional causes of poor quality may be added by operator, foreman, or quality-control person involved.

gram, as illustrated in Fig. 22-2. Also known as a fishbone diagram, it serves to remind those working on quality problems how various causes can operate to produce an eventual effect on the product manufactured.

It is considered good practice to maintain fishbone diagrams in the areas where the products are made. Oversize charts are sometimes displayed and workers are encouraged to add causes that may have been overlooked. The Cause-and-Effect chart is thus not only informative but also encourages participative effort on the part of those concerned with making the product.

Matrix Analysis

An example of a matrix analysis is given in Fig. 22-3. Four automatic polishers are in use in a particular manufacturing sector, and the frequency of defects is recorded. Instances where defects appear excessive are noted by circling the frequency.

While *DL* analysis could be applied, the defects identified initially are usually so excessive that people feel justified in simple judgment.

Defect Types	Auto-Polishers				Total	Mean
	No. 1	No. 2	No. 3	No. 4	Total	Mean
	2	2	0	0	4	1
Burns	⑩	0	0	6	16	4
Scratches	0	0	0	0	0	0
Nicks	0	4	0	⑧	12	3
Bent	⑫	0	0	0	12	3
Other	0	0	0	4	4	1
Total	24	6	0	18	48	
Mean	4	1	0	3		4.5

Fig. 22-3.   Matrix analysis. Illustration shows four automatic polishers and defects observed over a 10-day period. Problem areas are circled.

Later, when more refined decision making may be required, statistical analysis will become applicable.

Frequency Patterns

The nature and uses of frequency distribution patterns were presented early in this book, as a principal basis of quality-control

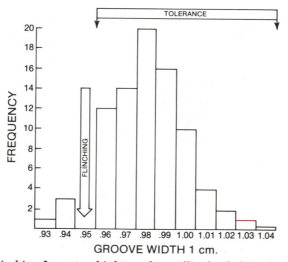

Fig. 22-4.   Flinching. Inspector shied away from calling borderline off-standard product. Instead, interval nearest tolerance limit is increased in frequency.

analysis. Since such patterns are easy to prepare and often reveal a wealth of information, quality circles make wide use of them. Some particularly interesting applications are given in Figs. 22-4 and 22-5.

Figure 22-4 shows dimensional measurements in relation to tolerances. The frequency pattern contains a "break" of no measurements near the lower tolerance. The reason for this, it was discovered, was inspector "flinching." That is, the inspector was reluctant to call borderline off-standard product as out-of-tolerance, not fully trusting his own measurements. Instead he placed these measurements into the next higher frequency, which is still within tolerances. This is a not uncommon tendency in inspection. Based on a frequency pat-

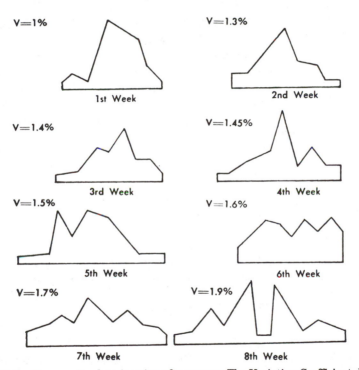

Fig. 22-5. Progressive deterioration of a process. The Variation Coefficient, V, continued to increase over the 8 weeks, after which maintenance brought equipment back to lower variations.

SOURCE: Enrick, N.L. *Management Control Manual for the Textile Industry*, 2nd ed. Melbourne, FL: Krieger Publ. Co., 1980. Used with permission.

tern analysis of inspection performance, the inspector can now be encouraged to record readings as seen.

Figure 22-5 presents a series of frequency patterns over a period of 8 weeks. The working area involved processing of a natural raw material containing seed, dirt, and fine particles that tended to clog in the equipment. As a result, the coefficient of variation V in percent rose from 1.0 to 1.9 during the 8 weeks. From a review of the frequency patterns it was decided that the current 8-week scouring and overhaul cycle on this equipment should be shortened to 4 weeks. In addition, Process Engineering was alerted to the need for a development program designed to (1) install exhaust ducts at key locations to lead off dust and other particles and (2) better protect bearings, gear trains, pulleys, and evener motions from grime, dirt, and grease effects.

In another instance, analysis of the frequency distribution of incoming components revealed that the vendor was screening out off-standard parts at the upper and lower ends of the distribution. This practice produced acceptable incoming quality, but obviously increased the cost of the product. Through communication with the vendor, encouragement was provided for a vendor reexamination of his own processing, with a view to reducing process variability. The eventual cost savings resulted in a lower purchase price.

While a wealth of information is often revealed by frequency patterns, quality circles must exercise care that conclusions are based on adequate data—such as from 50 to 100 data points per pattern. Otherwise, random fluctuations in build-up frequencies may lead to erroneous information and decisions.

Control Charts

It is only natural that control charts, the basic tool of quality control, find wide acceptance among quality circles. Abundant examples of the value of such charts have been presented. But a further illustration in Fig. 22-6 will be of interest. This is a control chart sequence in a spinning mill. Product is measured in unit weight terms, such as "grains per yard" in the drawing process. The sliver produced is next made into a preyarn known as "roving," which is stated in the reciprocal measure of "hank," that is yards per unit weight. Yarn count is a similar reciprocal measure, so that the higher the count, the thinner is the yarn.

It is apparent that heavy sliver in drawing produced heavy roving

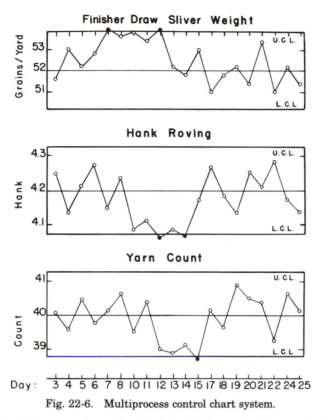

Fig. 22-6.   Multiprocess control chart system.

SOURCE: Enrick, N.L. *Management Control Manual for the Textile Industry*, 2nd ed. Melbourne, FL: Krieger Publ. Co., 1980. Used with permission.

and then heavy yarn. As a consequence, weight control requires the cooperative efforts of quality circles in the several departments affected. By maintaining more uniform weights several quality and cost benefits are obtained. For example, when light yarn is followed by heavy yarn in weaving, uneven fabric appearance results. Moreover, light fabrics incur customer claims and allowances, while heavy fabric is a give-away of costly materials.

Process Flow Charts

When brainstorming a quality or cost problem, it is often helpful to develop a process flow chart of the product, such as in Fig. 22-7. This is a simple type of process. Usually, assemblies will be more

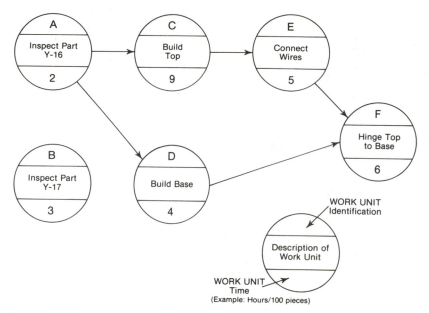

Fig. 22-7.   Flow process chart for a simple product. Numbers show work time in hours per 100 pieces.

complex, as represented by the product structure tree of Fig. 22-8 and the process flow chart of Fig. 22-9, for the multifrequency ringer of Fig. 22-10.

The purpose of process flow charts is to permit the group to keep in mind the sequence of operations, to recognize how individual assembly streams fit into the main stream, and to identify how a quality or cost problem at one stage may flow through the system and produce multiple effects downstream. For example, a loose fit at one stage may not cause immediate difficulties in assembly, but it may subsequently cause chatter and vibration in a subassembly.

Flow charts, similar to the process-to-process control chart presented above, are particularly useful in coordinating the quality, reliability, and cost efforts of several departments that are linked in the manufacture of a product.

Fault Tree Analysis

A fault tree is similar to a Cause-and-Effect diagram, but emphasizes reliability rather than quality failures. Various symbols

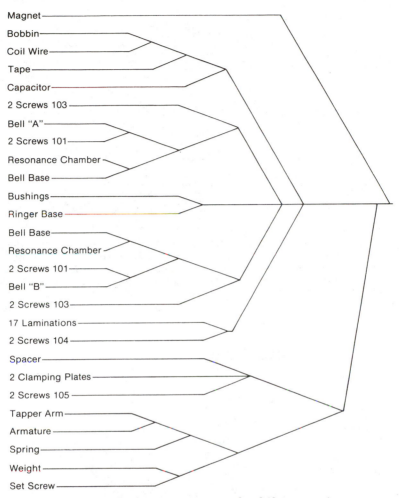

Fig. 22-8.   Product structure tree of multifrequency ringer.

are used, as illutrated in Fig. 22-11:

1. A simple box represents an event, or a fault, that is the result of several prior events. These prior events may represent a basic fault (circle) or a fault (four-sided diamond) which is not traced to further causes (but could be traced, if necessary).
2. An "or" gate notes that a subsequent failure may occur when

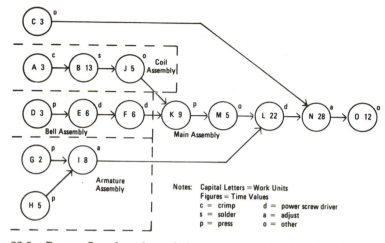

Fig. 22-9.   Process flow chart for multifrequency ringer. Explanation of operations: A = crimp and cut capacitor heads; B = solder capacitor to coil; C = insert rubber bushings; D = stake bells; E = attach bells to bases; F = attach bell bases to ringer base; G = rivet tapper arm to armature; H = press spring to armature; I = check and adjust armature; J = mount laminations; K = mount coil assembly; L = mount armature assembly; M = stake and magnetize magnet; N = adjust frequencies; O = pack.

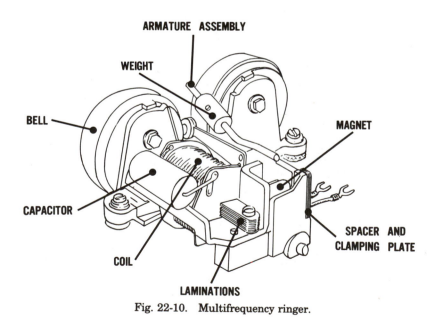

Fig. 22-10.   Multifrequency ringer.

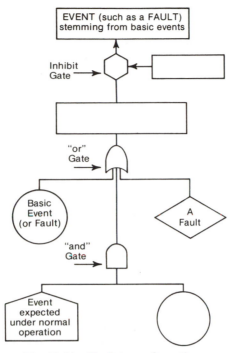

Fig. 22-11.   Fault-tree schematic.

any one of two (or more) prior events occurs. For example, computer failure may occur if (1) the program is incorrect *or* (2) the input is handled incorrectly.

3. An "and" gate indicates that subsequent failure occurs only if two or more events precede jointly. For example, power failure will occur if (1) the generator fails *and* (2) the automatic switch to auxiliary power does not work.

4. an "inhibit" gate is a conditional statement, similar to an "and." For example, power failure will occur if the generator fails, *given that* the switch to auxiliary power does not work.

5. Some events are expected under normal operation, such as the need to periodically replace parts on an airplane as a safety measure.

For a group working on product design or systems development, the fault tree is a guide to failure potentials. It is possible to evaluate overall reliability based on the reliabilities attached to each event,

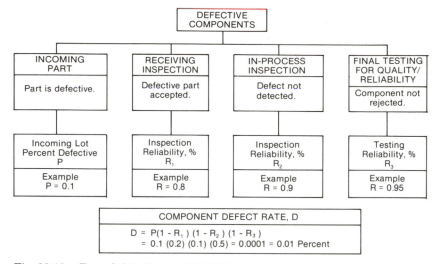

Fig. 22-12. Extended fault tree. Reliabilities *R* and proportion defective are added. *P* above is expressed as a decimal; that is, *P* of 0.1 represents 10 percent defectives.

as in Fig. 22-12. Although not shown specifically by an "and" gate, the final event "defective component" as a final product occurs only if all the following fail: incoming product quality, receiving inspection, in-process inspection, or final testing. The individual reliabilities *R* or, the incoming percent defective *P*, where applicable, are shown. Next these are combined in a manner familiar from Chapter 18, on Reliability Design, to yield a component defect rate of 0.01 percent. This is a low percentage for quality control, but may not be acceptable for reliability, since $R = 0.99$ or 99 percent, and a much higher value may be needed.

Failure Mode Analysis

Enhanced quality and reliability may also be obtainable from a review of failure modes. A matrix of processing factors and failure modes for an assembly operation is shown in Fig. 22-13. In cooperative endeavors, such as through quality circle work, the availability of a list of failure modes, failure causes, and failure factors at various processing stages should be helpful in tracing causes of failure and initiating corrective action. More important, the failure mode chart may serve as a problem prevention tool, by highlighting potential problem areas.

| POTENTIAL PROBLEM AREA | | FAILURE MODE | | | | |
PROCESS	FAILURE FACTOR	Intermittent	Short Circuit	Open Circuit	Value Change	Noise
				Failure Cause		
Wire-strip	Lead	Short	Long, broken	Short, nicked		Short
Solder	Temperature	Low		High, low	High	High, low
	Application	Thin	Thick	Thin		Thin, thick
Lead-cut and crimp	Tool	Dull, wrong		Dull, wrong	Dull, wrong	Dull, wrong
	Connection		Wrong			
	Force	Low		High, low		
Wire-wrap	Wire			Broken		
	Connection	Loose		Loose		Loose
Lead-bend	Stress on case	Excess	Excess		Excess	Excess
Wire-dress	Vibration sensitivity	Excess	Excess	Excess		Excess
	Residual stress	Excess	Excess	Excess		Excess

Fig. 22-13.   Failure mode matrix. Problem prevention, failure tracing, and correction of problems are facilitated by knowing failure causes, failure factors, and the processes involved.

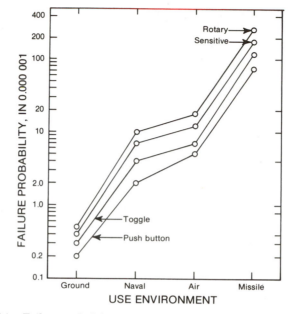

Fig. 22-14. Failure probabilities under varying stresses of use environment.

Use Environment

Failure analysis should also be use-cognizant. Different environments place varying stresses on equipments, and design and production must be planned accordingly. Figure 22-14 reviews four different types of switches under differing environments. A rotary switch, acceptable for ground or naval use, may thus be unsuitable for air or missile use. Thus reliability and quality are the result of constant effort, based on knowledge and good management.

Brainstorming

When a problem of quality, reliability, cost, productivity, or safety needs a combined effort for solution, brainstorming often helps in finding an avenue toward a solution. A brainstorming group may consist of production workers, their supervisor, and consultant process or design engineers from other departments, who have been requested to be of assistance.

All members are encouraged to input their ideas and recommendations in a disciplined, receptive environment. It is a cardinal rule

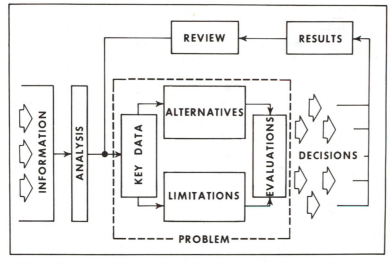

Fig. 22-15. Decision-implementation-and-review aspects of the idea-products of brainstorming.

that a new idea is not criticized. Instead, piggy-back type of contributions are encouraged: by building on an initial idea that may not have merit by itself, an original notion can gradually develop into a viable proposal.

When all ideas have been collected, and these have been further developed by orderly building ideas on the back of other ideas, it is time to evaluate and judge comparative merits. The alternatives receive evaluation in relation to the limitations that apply to each possible action. Decisions are formed. Eventually, those recommendations that find implementation will yield results that require further review and, possibly, additional brainstorming to deal with aspects not originally envisioned. A flow chart for this decision-implementation-and-review process appears in Fig. 22-15.

Brainstorming is a prime opportunity for creative participative management, using the type of analysis tools presented above.

Concluding Observations

This chapter has presented the principal, simple tools for analysis available particularly to Quality Circles as an aid in identifying and working on problems related to quality, reliability, cost, productivity, and other areas of concern. How to introduce and spread the quality circle concept is a topic that has received wide attention. The completely voluntary nature of the circle participation, the circle's control over prioritizing of problems, and the question of how cooperative and receptive to recommendations management is, are factors that determine how well the tools of analysis serve the circle and the organization as a whole. These are topics in human relations, motivation, and organization and management. One may hope that the success of quality circles in many firms may serve as an impetus for top management review of attitudes and policies toward a more participative style of decision-making and operation. Each such program must be adapted to the unique nature of each organization.

REVIEW QUESTIONS

1. Over a 1-month period, a shipping department observed these frequencies of mistakes:

Sent to wrong address	93
Wrong sizes or colors	60
Short	8
Damaged	15
Other	10

   Convert this information into a Pareto diagram.
2. Improved work performance is believed to be the result of the following:

   Type of worker: cooperative, responsible, motivated, able to get along well.

   Type of supervision: supervisor is a listener, good teacher, person of integrity, with leadership abilities.

   Workplace: well lighted, safe, comfortable, adequate space.

   Human relations: mutual respect, openness, fairness, considerateness, management sensitivity to worker needs.

   Show this information in a Cause-and-Effect diagram.
3. Using the matrix analysis given in Fig. 22-3, determine upper DL values.
4. Analysis of electrical resistance, in ohms, of an electronic component revealed these values:

X	f
20	10
21	25
22	50
23	20
24	18

The specification is 22 ± 2 ohm. Assuming this lot has just been received from your vendor, what would you conclude regarding his manufacturing and inspection process?

5. In what way can control charts for several successive processes aid in quality-control work?
6. What function does a process flow chart fulfill in brainstorming?
7. What is brainstorming?
8. What is flinching?
9. How can frequency distribution patterns help in the determination of maintenance frequencies?
10. A foundry has three casting ovens, experiencing pit marks, and chips, oversize and undersize problems in the output. What might a matrix analysis look like, assuming that pit marks are excessive in oven No. 1 and chips are excessive in oven No. 3?
11. The sugar and creamer set illustrated consists of the following:

> A tray with six spheres.
> A creamer with three spheres.
> A sugar cup with three spheres.
> The sugar cup has two handles; the creamer has one handle.
> The sugar and creamer are pressed from the same stainless steel disk.
> Lids are pressed from a separate disk, as is the metal tray.
> Each handle consists of two half-handles.
> Each sphere consists of two half-spheres.

Shapes of tray T, bowls B, and lids L are obtained by presswork-ing. Half-spheres HS and half-handles HH are purchased and then soldered on.

You are asked to develop a product structure tree and a process flow chart. Note that different people may design different proc-essing arrangements with resultant variations in process flow.

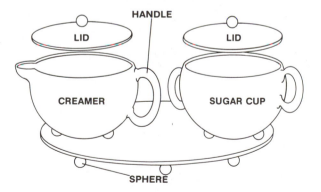

Appendixes

APPENDIX 1

METHODS OF COMPUTING STANDARD DEVIATIONS

For most practical purposes, the standard deviation of a production lot or process may be estimated by use of the method of average ranges shown in this book. There are, however, occasions when it is desired to make more precise determinations, using direct computations. Several methods are available for this purpose, of which the most commonly used are described here. They have been placed here, in the Appendix, since they are applied less frequently than the range methods in quality control work.

Basic Method

The basic method of computing standard deviations is shown in Appendix 1A. The procedure illustrated by Fig. 6-1, page 70, represents an expansion of the basic method, for use in frequency distributions.

APPENDIX 1A

BASIC METHOD OF COMPUTING STANDARD DEVIATION

(Example of Tensile Strength Determinations)

Tensile Strength, Pounds	Average	Deviation from Average	Deviation from Average, Squared
(X)	$(\bar{X})$	$(X - \bar{X})$	$(X - \bar{X})^2$
64	62	+2	4
60	62	−2	4
62	62	0	0
66	62	+4	16
58	62	−4	16
Totals 310	—	0	40

1. Average, $\bar{X} = \dfrac{\Sigma X}{N} = \dfrac{310}{5} = 62$*

2. Variance, $\sigma^2 = \Sigma(X-\bar{X})^2/N = 40/5 = 8$

3. Standard Deviation, $\sigma = \sqrt{\sigma^2} = \sqrt{8} = 2.828$

*Σ = "Sum of," N = Sample Size.

Unbiased Method

When it is desired to estimate the standard deviation of a production lot or process from a small sample, the basic method shown in Appendix 1A usually underestimates the true standard deviation slightly. In order to correct for this bias, the sample size N in the formula for the variance is replaced by N-1. The latter expression is called "Degrees of Freedom." Appendix 1B shows the method. Obviously, as the sample size becomes larger, the difference between N and N-1 becomes less important. For this reason, many texts use simply N in place of N-1 when the sample size exceeds 25 or 30. This was also done in Fig. 6-1, page 70.

APPENDIX 1B

ESTIMATING STANDARD DEVIATION OF A LOT FROM A SAMPLE

(Example of Tensile Strength Determinations)

Tensile Strength, Pounds	Average	Deviation from Average	Deviation from Average, Squared
(X)	$(\overline{X})$	$(X - \overline{X})$	$(X - \overline{X})^2$
64	62	+2	4
60	62	−2	4
62	62	0	0
66	62	+4	16
58	62	−4	16
Totals 310	—	0	40

1. Average, $\overline{X}, = \dfrac{\Sigma X}{N} = \dfrac{310}{5} = 62$

2. Variance, $\sigma^2 = \Sigma(X-\overline{X})^2/(N-1) = 40/4 = 10$

3. Standard Deviation, $\sigma = \sqrt{\sigma^2} = \sqrt{10} = 3.162$

Short-Cut Method

When a desk calculator or an electronic computer is available, it is preferable to use the short-cut method demonstrated in Appendix 1C. The answer should be identical to that found by the method of Appendix 1B.

APPENDIX 1C

ESTIMATING STANDARD DEVIATION OF A PRODUCTION LOT FROM A SAMPLE—SHORT-CUT METHOD OF COMPUTATION

(Example of Tensile Strength Determinations)

Tensile Strength, Pounds	Strength Squared
(X)	$(X)^2$
64	4,096
60	3,600
62	3,844
66	4,356
58	3,364
Totals 310	19,260

1. $(\Sigma X)^2 = (310)^2 = 96,100$

2. Average, $\overline{X} = \dfrac{\Sigma X}{N} = \dfrac{310}{5} = 62$

3. $N = 5; N - 1 = 4; N(N-1) = 20$

4. Variance, $\sigma^2 = \dfrac{\Sigma(X)^2}{N-1} - \dfrac{(\Sigma X)^2}{N(N-1)}$

$= \dfrac{19,260}{4} - \dfrac{96,100}{20}$

$= 4,815 - 4,805 = 10$

5. Standard Deviation, $\sigma = \sqrt{\sigma^2} = \sqrt{10} = 3.162$

Grouped Data

When data are grouped in the form of a frequency distribution, the computations by means of the short-cut and unbiased method proceed as illustrated in Appendix 1D. This is the fastest procedure for grouped data.

APPENDIX 1D

ESTIMATING STANDARD DEVIATION OF A LOT FROM A SAMPLE—SHORT-CUT METHOD OF COMPUTATION FOR GROUPED DATA

(Example of Product Weight)

Product Weight (oz.)	Midvalue (oz.)	Frequency (f)	Deviation* (d)	f × d = fd	fd × d = fd²
1.1-1.19	1.15	2	−3	−6	18
1.2-1.29	1.25	2	−2	−4	8
1.3-1.39	1.35	4	−1	−4	4
1.4-1.49	1.45	6	0	0	0
1.5-1.59	1.55	3	+1	+3	3
1.6-1.69	1.65	1	+2	+2	4
1.7-1.80	1.75	2	+3	+6	18
Totals	−	20		−3	55

1. Average, $\overline{X} = A + \dfrac{\Sigma(fd) \times i}{N} = 1.45 + \dfrac{-3 \times 0.1}{20} = 1.435$

2. Variance, $\sigma^2 = \left[\dfrac{\Sigma fd^2}{N-1} - \dfrac{(\Sigma fd)^2}{N(N-1)} \right] \times i^2$

$$- \left[\dfrac{55}{19} - \dfrac{(-3)^2}{20 \times 19} \right] \times 0.01 = 0.0287$$

3. Standard Deviation $= \sqrt{\sigma^2} = \sqrt{0.0287} = 0.169$

Three-Factor Analysis

Our experiment in Chapter 19, investigating the effect on flexure life (in 100s of cycles) of strandings, slacks, and sizes, yields data from which the standard deviations for main effects, interaction effects, and

---

*Deviations are measured in full class interval units, from a tentatively assumed average, A. In this example, A was taken at 1.45. The class interval units, i, are the differences between successive midvalues, representing 0.1 oz. each. For example, 1.25 − 1.15 = 0.1.

experimental error can be calculated as shown in Appendix 1F; it represents a three-factor replicated analysis. The overall (or total) standard deviation is obtained by means of the Pythagorean or vectorial addition discussed in Chapter 12. The individual variances can be expressed as percents of the overall variance, as in Fig. A1-F. From this it is apparent that the effect of strands on life quality is relatively small. On the other hand, sizes, slacks, and interactions demonstrate some magnitude and, thus, importance. The relatively large error (greater than either slacks or interactions) suggests that factors of variation are operating which have not yet been pinpointed. Future experiments may now be designed to investigate these factors as part of the continuing efforts to enhance the reliability of the product.

Some readers may note the slight difference between the error standard deviation of 2.8 obtained by the mathematical methods of Appendix 1F and the value of 2.7 given in Chapter 19 as a result of the quicker range method. The somewhat lesser precision of the range method was the price we paid for the simplified range approach. Both methods are merely estimates of the true (but unknown) experimental error term (which is the value obtainable with theoretically infinite sampling).

Multiple Classification

When data are classified with regard to source, the standard deviation ascribable to each may be estimated. For example, for the percent defectives on three drop hammers, in Chapter 2, we can calculate the standard deviation for the two main effects (machines and lot numbers). A remaining residual variation, labeled "error" in the sense that it cannot be ascribed to a specific factor, is then combined with the main effects to yield overall variation. We restate the percentages, omitting lots Nos. 3 and 8 that represented non-normal output:

Lot No.	Machines Hammer No. 1	Hammer No. 2	Hammer No. 3	Lot Average	Lot Total
1	2.0	0.8	1.9	1.57	4.7
2	1.8	1.3	2.2	1.77	5.3
4	1.1	0.5	0.8	0.80	2.4
5	1.3	1.2	1.2	1.23	3.7
6	1.8	1.2	1.2	1.40	4.2
7	0.9	1.4	0.8	1.03	3.1
Machine Average	1.48	1.07	1.35	1.30	
Machine Total	8.9	6.4	8.1		23.4

This two-way model can readily grow to three classifications. The further factor "suppliers" would be introduced if the first three lots had been made from materials provided by source A, while the second three lots came from B. We shall confine our analysis to the two-factor situation, by entering the data above in the self-explanatory system in Appendix 1E.

The analysis deals with the two main factors (variation among lots and variation among machines), the error variation, and the total or overall variation. "Error" refers to those variations not ascribable to a factor in the analysis. It is akin to chance fluctuations and also known as "residual" variation.

In this particular example, the variations between lots and between machines are each smaller than the error variation (see lines k and l), meaning that there are no significant differences among lots or among machines. A formal test of significance, using F-ratios along the lines discussed in Chapter 12, is possible.

APPENDIX 1E

ESTIMATING STANDARD DEVIATION FOR A TWO-FACTOR PROBLEM

(Example of Drop Hammer Machines from the preceding page.) Main factors under study are (1) lots and (2) machines. Combining these with the residual or error variation, we obtain total variation for the data.

Sources of Variation

Computation Steps	(1) Between Lots	(2) Between Machines	(3) Error	(4) Total
a. Enter Squared Totals	4.7^3 $+5.3^2$ etc., for all 6	8.9^2 $+6.4^2$ $+8.1^2$	. . .	2.0^2 $+0.8^2$ etc., for all 18 observations
b. Sum of Squares in (a)	96.88	185.78	. . .	34.22
c. No. of Observations for Each Square in (a)	3	6	. . .	1
d. Divide: b/c	32.293	30.963	. . .	34.220
e. Correction Factor*	30.420	30.420	. . .	30.420
f. Formula for Sum of Squares	$d_1 - e$	$d_2 - e$	$g_4 - g_{1,2}$	$d_4 - e$
g. Sum of Squares	1.873	0.543	1.384	3.800
h. Degrees of Freedom, DF**	$6 - 1 = 5$	$3 - 1 = 2$	10	$18 - 1 = 17$
i. Mean Square: g/h	0.375	0.272	0.138	. . .
j. Formula for Estimated Variance	$(i_1 - i_3)/6$	$(i_2 - i_3)/6$	i_3	$k_1 + k_2 + k_3$
k. Estimated Variance	$= 0.0395$	0.0223	0.138	1.988
l. Standard Deviation, $\sqrt{k}$	0.1987	0.149	0.371	0.446†

*Correction Factor = (Grand Total)$^2/n$ = $23.4^2/18$ = 30.42 where N is the total number of observations.

**For the 6 lots, $DF = 6 - 1 = 5$; for the 3 machines, $DF = 3 - 1 = 2$, for the 18 tests in all $DF = 18 - 1 = 17$. Finally, the DF-total of 17 minus DF-lots of 5 minus DF-machines of $2 = 17 - 5 - 2 = 10$ DF for the error variation.

†Individual standard deviations do not sum to a simple total. Instead the Pythagorean addition of Chapter 12 applies.

APPENDIX 1F

ESTIMATING STANDARD DEVIATION FOR A THREE-WAY REPLICATED EXPERIMENT

Computation Steps	Factors			Interactions				(8) Error	(9) Total
	(1) Strands	(2) Slacks	(3) Sizes	(4) Strand × Slack	(5) Strand × Size	(6) Slack × Size	(7) Three-Way		
a. Enter Squared Totals	266^2 315^2 etc.	96^2 105^2 etc.	134^2 255^2 etc.	43^2 40^2 etc.	73^2 113^2 etc.	12^2 14^2 etc.	6^2 7^2 etc.	— — —	2^2 4^2 etc.
b. Sum Squares in (a)	169,981	71,263	119,845	36,271	60,927	25,685	13,277	—	6,865
c. No. of Observations per Each Square in (a)	30	12	20	6	10	4	2	—	1
d. Divide: b/c	5,666	5,939	5,992	6,045	6,093	6,421	6,638	—	6,865
e. From d Subtract Correction Factor*	40	313	366	419	467	795	1,012	—	1,239
f. Sum of Squares	$e_1 = 40$	$e_2 = 313$	$e_3 = 366$	$e_4 - f_{1,2} = 66$	$e_5 - f_{1,3} = 61$	$e_6 - f_{2,3} = 116$	$e_7 - f_{1 \to 6} = 50$	$e_9 - d_7 = 277$	$e_9 = 1,239$
g. Degrees Freedom, DF	$2-1=1$	$5-1=4$	$3-1=2$	$1 \times 4 = 4$	$1 \times 2 = 2$	$4 \times 2 = 8$	$1 \times 4 \times 2 = 8$	$30(2-1) = 30$	$60-1 = 59$
h. Mean Square, f/g	40	78.25	185	16.5	30.5	14.5	6.25	7.57	—
i. Estimated Variance for Factors and Interactions	$\frac{h_1 - h_8}{c_1} = 1.1$	$\frac{h_2 - h_8}{c_2} = 5.9$	$\cdots$ 8.8	and similarly 1.5	$\cdots$ 2.3	$\frac{h_6 - h_8}{c_6}$ 1.7	$\cdots$ 0^{**}	h_8 7.57	Row sum 28.9
j. Individual Variances as a Percent of Total Variance	3.8	20.4	30.4	5.2	8.0	5.9	0	26.2	100
k. Standard Deviation, $\sqrt{i}$	1.0	2.4	2.9	1.2	1.5	1.3	0	2.8	5.4

*Correction Factor = (Grand Total)²/(No. of Observations) = 581²/60 = 5626.

**Set negatives equal to zero.

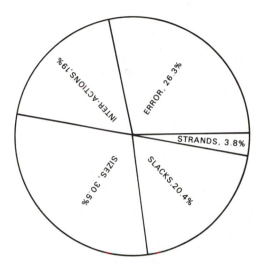

Fig. A1-F. Relative magnitude of estimated net components of variance for three-factor replicated experiment. Total variance is set at 100 percent.

APPENDIX 2

OPERATING CHARACTERISTICS FOR LOT-BY-LOT SAMPLING PLANS

This Appendix brings the operating characteristic (OC) curves for the sets of master plans for sequential and single sampling in Tables 2-1 and 2-2. The curves show the risks of sampling error corresponding to the lot qualities which may be encountered in practice.

Although it is usually simpler to look up key values of sampling risks in tabular form (Chapter 3, Table 3-2, covering all sampling plans in Tables 2-1 and 2-2), only a curve can indicate all of the risks for a given sampling plan.

A particular situation requiring OC curve application occurs when the average level of a process, in percent defective, is to be estimated from past experience with a series of sampling acceptances and rejections. For example, a production process has been controlled through lot-by-lot sampling, using an Acceptable Quality Level (AQL) of 1 percent for lot sizes between 500 and 799. Over the past month, there were 80 percent acceptances and 20 percent rejections. Estimation of the process average in terms of percent defectives for the period under study proceeds as follows:

1. Refer to the curve for lot sizes 500 to 799, AQL of 1 percent, in Fig. A2-1.
2. Entering the "Probability of Acceptance" scale at the level of 80 percent, move horizontally until the curve for AQL = 1 is reached.
3. Moving down vertically, we find on the base scale the estimated process average of 2 percent.

This quick and ready method has a potential bias: when there is a large number of unusually bad lots, the estimate tends to understate the percent defective. An alternate approach which avoids this bias was given in Chapter 2, in the section "Long Term Value of Lot-by-Lot Inspection," using the relation:

$$\text{Process Average, Percent} = \frac{\text{Total No. of Defectives Found}}{\text{Total No. of Units Sampled}}$$

Usually this method involves more record-keeping, and it may give undue weight to occasional, isolated, very bad lots. Some plants find it useful to compare the results of the OC-curve read-out with the formula result:

(Continued on page 361)

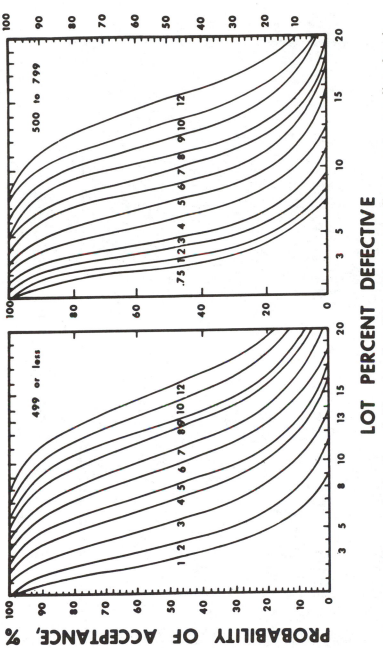

LOT PERCENT DEFECTIVE

Fig. A2-1. OC curves for lot sizes of 499 or less than 500 to 799, corresponding to master sampling plans in Tables 2-1 and 2-2. Figures next to each curve represent Acceptable Quality Level (AQL) in percent.

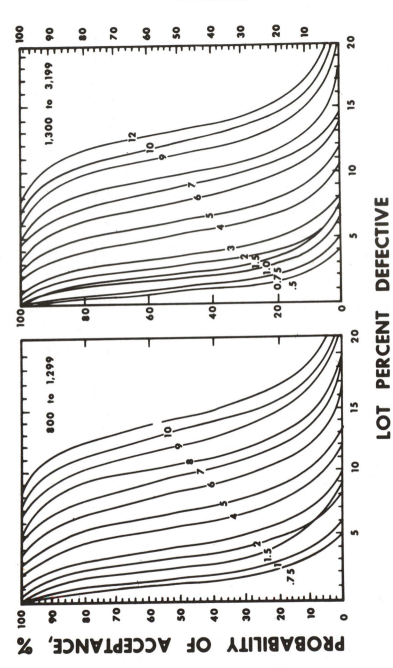

Fig. A2-2. OC curves for lot sizes 800 to 1299 and 1300 to 3199 corresponding to master sampling plans in Tables 2-1 and 2-2. Figures next to each curve show AQL.

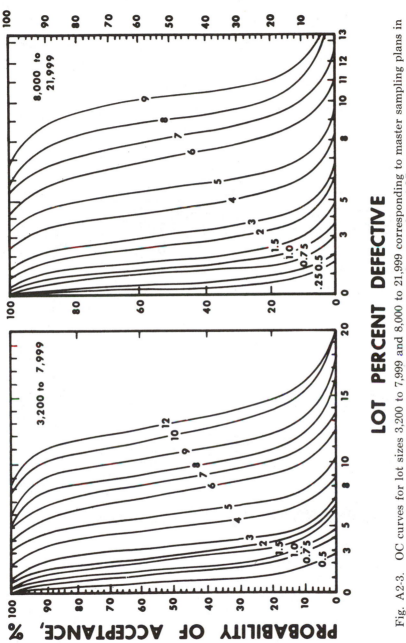

Fig. A2-3. OC curves for lot sizes 3,200 to 7,999 and 8,000 to 21,999 corresponding to master sampling plans in Tables 2-1 and 2-2. Figures next to each curve show AQL.

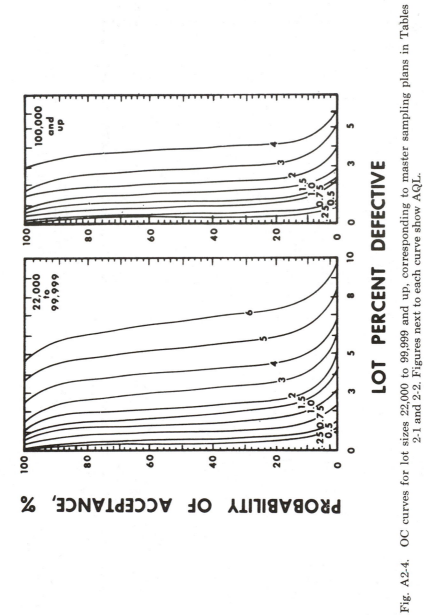

Fig. A2-4. OC curves for lot sizes 22,000 to 99,999 and up, corresponding to master sampling plans in Tables 2-1 and 2-2. Figures next to each curve show AQL.

(Continued from page 356)

$$\text{Ratio of Process Averages} = \frac{\text{Process Average from OC-Curves}}{\text{Process Average from Formula}}$$

A high ratio (from 1.6 or greater) indicates relatively frequent occurrence of individual lots with excessively high percent defective.[*]

Other uses of OC curves were given in Chapter 3. Each of the approximately 70 curves in Figs. A2-1 to A2-4 applies equally to either a sequential or single sampling plan for the given lot sizes and AQL's. As noted in Chapter 3, with increasing lot sizes and correspondingly larger samples, the OC curves become steeper, representing sharper discrimination between good and poor quality.

---

[*]Management uses of process averages for the purpose of quality history accumulation and corrective engineering and production feedback are further discussed in R. H. Lester, N. L. Enrick, and H. E. Mottley, *Quality Control for Profit* (New York: Industrial Press, 1977).

APPENDIX 3

REFERENCE TABLES AND CHARTS

Certain tabular values, such as for areas under the normal curve or for control chart calculations, will be needed not just in connection with the procedures of the chapter in which they were first introduced, but again in subsequent chapters. For the reader's convenience, these tabular values are collected in this appendix. Moreover, when first presented in each chapter, tables have often been abbreviated for the purpose of simplified presentation (covering only the relatively more frequent application needs). The tables that follow have been expanded to be complete for the entire range of possible requirements.

Furthermore, a Weibull chart is presented as a means of facilitating applications. The illustrative filled-out example in Chapter 17 will serve the purpose of demonstrating the use of the chart, but a larger scale replica is more suitable for applications.

The Normal Curve table is taken from Norbert L. Enrick, *Management Handbook of Decision Oriented Statistics*, Melbourne, FL: Krieger Publ. Co., 1980, pp. 225–227.

APPENDIX 3A

DETAILED TABLE OF AREAS UNDER THE NORMAL CURVE

Range of std. dev.	Items within std. dev. range, %	Range of std. dev.	Items within std. dev. range, %	Range of std. dev.	Items within std. dev. range, %	Range of std. dev.	Items within std. dev. range, %
0.01	0.399	0.42	16.276	0.83	29.673	1.24	39.251
0.02	0.798	0.43	16.640	0.84	29.955	1.25	39.435
0.03	1.197	0.44	17.003	0.85	30.234	1.26	39.617
0.04	1.595	0.45	17.364	0.86	30.511	1.27	39.796
0.05	1.994	0.46	17.724	0.87	30.785	1.28	39.973
0.06	2.392	0.47	18.082	0.88	31.057	1.29	40.147
0.07	2.790	0.48	18.439	0.89	31.327	1.30	40.320
0.08	3.188	0.49	18.793	0.90	31.594	1.31	40.490
0.09	3.586	0.50	19.146	0.91	31.859	1.32	40.658
0.10	3.983	0.51	19.497	0.92	32.121	1.33	40.824
0.11	4.380	0.52	19.847	0.93	32.381	1.34	40.988
0.12	4.776	0.53	20.194	0.94	32.639	1.35	41.149
0.13	5.172	0.54	20.540	0.95	32.894	1.36	41.309
0.14	5.567	0.55	20.884	0.96	33.147	1.37	41.466
0.15	5.962	0.56	21.226	0.97	33.398	1.38	41.621
0.16	6.356	0.57	21.566	0.98	33.646	1.39	41.774
0.17	6.749	0.58	21.904	0.99	33.891	1.40	41.924
0.18	7.142	0.59	22.240	1.00	34.134	1.41	42.073
0.19	7.535	0.60	22.575	1.01	34.375	1.42	42.220
0.20	7.926	0.61	22.907	1.02	34.614	1.43	42.364
0.21	8.317	0.62	23.237	1.03	34.850	1.44	42.507
0.22	8.706	0.63	23.565	1.04	35.083	1.45	42.647
0.23	9.095	0.64	23.891	1.05	35.314	1.46	42.786
0.24	9.483	0.65	24.215	1.06	35.543	1.47	42.922
0.25	9.871	0.66	24.537	1.07	35.769	1.48	43.056
0.26	10.257	0.67	24.857	1.08	35.993	1.49	43.189
0.27	10.642	0.68	25.175	1.09	36.214	1.50	43.319
0.28	11.026	0.69	25.490	1.10	36.433	1.51	43.448
0.29	11.409	0.70	25.804	1.11	36.650	1.52	43.574
0.30	11.791	0.71	26.115	1.12	36.864	1.53	43.699
0.31	12.172	0.72	26.424	1.13	37.076	1.54	43.822
0.32	12.552	0.73	26.730	1.14	37.286	1.55	43.943
0.33	12.930	0.74	27.035	1.15	37.493	1.56	44.062
0.34	13.307	0.75	27.337	1.16	37.698	1.57	44.179
0.35	13.683	0.76	27.637	1.17	37.900	1.58	44.295
0.36	14.058	0.77	27.935	1.18	38.100	1.59	44.408
0.37	14.431	0.78	28.230	1.19	38.298	1.60	44.520
0.38	14.803	0.79	28.524	1.20	38.493	1.61	44.630
0.39	15.173	0.80	28.814	1.21	38.686	1.62	44.738
0.40	15.542	0.81	29.103	1.22	38.877	1.63	44.845
0.41	15.910	0.82	29.389	1.23	39.065	1.64	44.950

(continued)

Range of std. dev.	Items within std. dev. range, %	Range of std. dev.	Items within std. dev. range, %	Range of std. dev.	Items within std. dev. range, %	Range of std. dev.	Items within std. dev. range, %
1.65	45.053	2.06	48.030	2.47	49.324	2.88	49.801
1.66	45.154	2.07	48.077	2.48	49.343	2.89	49.807
1.67	45.254	2.08	48.124	2.49	49.361	2.90	49.813
1.68	45.352	2.09	48.169	2.50	49.379	2.91	49.819
1.69	45.449	2.10	48.214	2.51	49.396	2.92	49.825
1.70	45.543	2.11	48.257	2.52	49.413	2.93	49.831
1.71	45.637	2.12	48.300	2.53	49.430	2.94	49.836
1.72	45.728	2.13	48.341	2.54	49.446	2.95	49.841
1.73	45.818	2.14	48.382	2.55	49.461	2.96	49.846
1.74	45.907	2.15	48.422	2.56	49.477	2.97	49.851
1.75	45.994	2.16	48.461	2.57	49.492	2.98	49.856
1.76	46.080	2.17	48.500	2.58	49.506	2.99	49.861
1.77	46.164	2.18	48.537	2.59	49.520	3.00	49.865
1.78	46.246	2.19	48.574	2.60	49.534	3.01	49.869
1.79	46.327	2.20	48.610	2.61	49.547	3.02	49.874
1.80	46.407	2.21	48.645	2.62	49.560	3.03	49.878
1.81	46.485	2.22	48.679	2.63	49.573	3.04	49.882
1.82	46.562	2.23	48.713	2.64	49.585	3.05	49.886
1.83	46.638	2.24	48.745	2.65	49.598	3.06	49.889
1.84	46.712	2.25	48.778	2.66	49.609	3.07	49.893
1.85	46.784	2.26	48.809	2.67	49.621	3.08	49.897
1.86	46.856	2.27	48.840	2.68	49.632	3.09	49.900
1.87	46.926	2.28	48.870	2.69	49.643	3.10	49.903
1.88	46.995	2.29	48.899	2.70	49.653	3.11	49.906
1.89	47.062	2.30	48.928	2.71	49.664	3.12	49.910
1.90	47.128	2.31	48.956	2.72	49.674	3.13	49.913
1.91	47.193	2.32	48.983	2.73	49.683	3.14	49.916
1.92	47.257	2.33	49.010	2.74	49.693	3.15	49.918
1.93	47.320	2.34	49.036	2.75	49.702	3.16	49.921
1.94	47.381	2.35	49.061	2.76	49.711	3.17	49.924
1.95	47.441	2.36	49.086	2.77	49.720	3.18	49.926
1.96	47.500	2.37	49.111	2.78	49.728	3.19	49.929
1.97	47.558	2.38	49.134	2.79	49.736	3.20	49.931
1.98	47.615	2.39	49.158	2.80	49.744	3.21	49.934
1.99	47.670	2.40	49.180	2.81	49.752	3.22	49.936
2.00	47.725	2.41	49.202	2.82	49.760	3.23	49.938
2.01	47.778	2.42	49.224	2.83	49.767	3.24	49.940
2.02	47.831	2.43	49.245	2.84	49.774	3.25	49.942
2.03	47.882	2.44	49.266	2.85	49.781	3.26	49.944
2.04	47.932	2.45	49.286	2.86	49.788	3.27	49.946
2.05	47.982	2.46	49.305	2.87	49.795	3.28	49.948

(*continued*)

Range of std. dev.	Items within std. dev. range, %	Range of std. dev.	Items within std. dev. range, %	Range of std. dev.	Items within std. dev. range, %	Range of std. dev.	Items within std. dev. range, %
3.29	49.950	3.47	49.974	3.65	49.987	3.83	49.994
3.30	49.952	3.48	49.975	3.66	49.987	3.84	49.994
3.31	49.953	3.49	49.976	3.67	49.988	3.85	49.994
3.32	49.955	3.50	49.977	3.68	49.988	3.86	49.994
3.33	49.957	3.51	49.978	3.69	49.989	3.87	49.995
3.34	49.958	3.52	49.978	3.70	49.989	3.88	49.995
3.35	49.960	3.53	49.979	3.71	49.990	3.89	49.995
3.36	49.961	3.54	49.980	3.72	49.990	3.90	49.995
3.37	49.962	3.55	49.981	3.73	49.990	3.91	49.995
3.38	49.964	3.56	49.981	3.74	49.991	3.92	49.996
3.39	49.965	3.57	49.982	3.75	49.991	3.93	49.996
3.40	49.966	3.58	49.983	3.76	49.992	3.94	49.996
3.41	49.968	3.59	49.983	3.77	49.992	3.95	49.996
3.42	49.969	3.60	49.984	3.78	49.992	3.96	49.996
3.43	49.970	3.61	49.985	3.79	49.992	3.97	49.996
3.44	49.971	3.62	49.985	3.80	49.993	3.98	49.997
3.45	49.972	3.63	49.986	3.81	49.993	3.99	49.997
3.46	49.973	3.64	49.986	3.82	49.993	4.00	49.998

APPENDIX 3B

CONTROL LIMIT AND STANDARD DEVIATION CALCULATION FACTORS

Sample Size (No. of Observations per Sample) n	Factors for Control Charts for Averages			Factors for Control Charts for Ranges		Factor for Standard Deviation F_r
	Based on Specifications, F_s	Based on Center Line		Upper Control Limit D_4	Lower Control Limit D_3	
		A_2	$\frac{2}{3}A_2$			
2	1.49	1.880	1.253	3.267	0	0.8865
3	1.14	1.023	0.682	2.575	0	0.5907
4	1.02	0.729	0.486	2.282	0	0.4857
5	0.94	0.577	0.385	2.115	0	0.4299
6	0.90	0.483	0.322	2.004	0	0.3946
7	0.84	0.419	0.279	1.924	0.076	0.3698
8	0.80	0.373	0.249	1.864	0.136	0.3512
9	0.82	0.337	0.225	1.816	0.184	0.3367
10	0.80	0.308	0.205	1.777	0.223	0.3249
11	0.78	0.285	0.190	1.744	0.256	0.3152
12	0.77	0.266	0.177	1.716	0.284	0.3069
13	0.76	0.249	0.166	1.692	0.308	0.2998
14	0.75	0.235	0.157	1.671	0.329	0.2935
15	0.74	0.223	0.149	1.652	0.348	0.2880
16	0.73	0.212	0.141	1.636	0.364	0.2831
17	0.73	0.203	0.135	1.621	0.379	0.2787
18	0.72	0.194	0.129	1.608	0.392	0.2747
19	0.72	0.187	0.125	1.596	0.404	0.2711
20	0.71	0.180	0.120	1.586	0.414	0.2677
21	0.70	0.173	0.115	1.575	0.425	0.2647
22	0.70	0.167	0.111	1.566	0.434	0.2618
23	0.70	0.162	0.108	1.557	0.443	0.2592
24	0.69	0.157	0.105	1.548	0.452	0.2567
25	0.68	0.153	0.102	1.541	0.459	0.2544

APPENDIX 3C

WEIBULL CHART

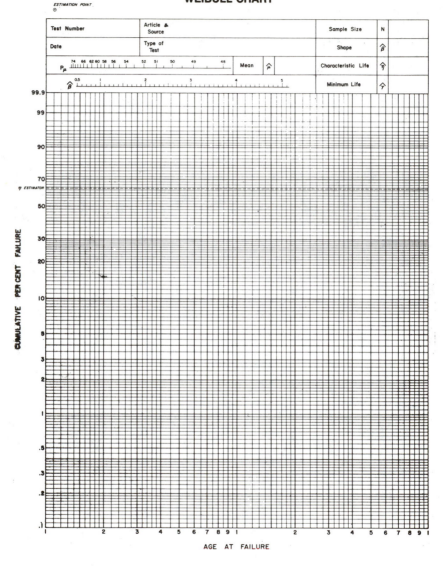

ESTIMATION POINT

| Test Number | Article & Source | | Sample Size | N | |
| Date | Type of Test | | Shape | $\hat{\beta}$ | |

Mean $\hat{\mu}$ Characteristic Life $\hat{\eta}$

Minimum Life $\hat{\gamma}$

CUMULATIVE PER CENT FAILURE

AGE AT FAILURE

APPENDIX 3D

FACTORS FOR CONTROL LIMITS AND PROCESS STANDARD DEVIATION BASED ON THE AVERAGE $\bar{s}$ OF A SERIES OF SAMPLE STANDARD DEVIATIONS s

Background

Control charts and related values are generally calculated on the basis of sample average $\bar{X}$ and sample range R. Automated gaging and computer-assisted measurement, however, make it practical to substitute the somewhat more precise sample standard deviation s for R. Noting that DF = Degrees of Freedom = $n - 1$ in the calculation of the variance, then for a sample of n measurements, each of value X_i, we find:

Sample variance, $s^2 = \Sigma(X - \bar{X})/(n-1)$
Sample standard deviation, $s = \sqrt{s^2}$

Example of $\bar{s}$

Assume $k = 4$ samples each of size $n = 2$, with these results in 0.0001 cm of measured diameter of a cylindrical part:

	12	13	11	14
	16	15	13	18
$\bar{X}$	14	14	12	16
s^2	8	2	2	8
s	2.8	1.4	1.4	2.8

where the first s^2, as an illustration, is $[(12 - 14)^2 + (16 - 14)^2]/(2 - 1)$ or $(2^2 + 2^2)/(2-1) = 8$. Next,

$$\bar{s} = (2.8 + 1.4 + 1.4 + 2.8)/4 = 2.1$$

where $k = 4$ is the number of samples in the series.

Calculation of Control Limits

Control Limits are found in the usual manner, excepting that the new values for the applicable factors are used, as separately tabulated at the end in this Appendix.

Control Limits Calculated from Specification Limits

$$UCL = \text{Upper Specification Limit} - F_{sd} \times \bar{s}$$
$$LCL = \text{Lower Specification Limit} + F_{sd} \times \bar{s}$$

For our example, with $\bar{s} = 2.1$, assuming a specification of 14 ± 6, we find:

$$UCL = 20 - 2.12(2.1) = 15.548$$
$$LCL = 8 + 2.12(2.1) = 12.45$$

Note that one sample average at $\bar{X} = 12$ is below the LCL.

Control Limits Based on Center Lines

$$\begin{aligned}
CL(99.7\% \text{ Confidence}) &= \text{Process Average} \pm A_3 \times 2.1 \\
&= 14 \pm 2.659(2.1) \\
&= 14 \pm 5.6
\end{aligned}$$

$$\begin{aligned}
CL(95\% \text{ Confidence}) &= \text{Process Average} \pm \tfrac{2}{3}A_3 \times 2.1 \\
&= 14 \pm 1.773(2.1) \\
&= 14 \pm 3.7
\end{aligned}$$

Note that none of the sample averages is outside control. Thus the production process is in a state of statistical control; however, the specification limits are closer than the process capability, and hence the specification-based control limits do show one out-of-control point. It has previously been emphasized that control limits should not be calculated, nor are control charts applicable, where process capability does not meet tolerances established for the product made.

Control Limits for Standard Deviation s

$$UCL = B_4 \times \bar{s}$$
$$LCL = B_3 \times \bar{s}$$

For our example, using $\bar{s} = 2.1$ and $n = 2$, we find:

$$UCL = 3.267(2.1) = 6.86$$
$$LCL = 0(2.1) = 0$$

Process Standard Deviation

Previously we noted that an unbiased estimate of the process standard deviation is obtained from a sample by using DF instead of n. More sophisticated considerations, however, reveal that some degree of underestimation may still remain when n is very small. In order to allow for this underestimate, we divide the observed $\bar{s}$ by the factor C_4. For example,

$$\begin{aligned}
\text{Population Standard Deviation} &= \bar{s}/C_4 \\
&= 2.1/.7979 \\
&= 2.632
\end{aligned}$$

Concluding Observations

In practice, the use of R has the advantage of simple calculation and immediate understanding of what this represents. R is particularly well suited for pencil calculations and direct control chart plotting. On the other hand, s is a somewhat more precise measurement, since it uses *all* of the values measured. Only when $n = 2$ does R measure all of the values directly. Some computerized programs, involving automated gaging or control, thus apply s rather than R. The factors used then must be modified, as shown by the tabulation that follows.

It will be noted that there is no change in theory, concepts, or methodology of statistical quality control in the use of $\bar{s}$ in place of $\bar{R}$. Merely the availability of computerized calculations permits the use of a somewhat (10–15 percent) more precise measure, the estimated standard deviation in place of the average range.

APPENDIX 3 D

CONTROL LIMIT AND STANDARD DEVIATION FACTORS BASED ON AVERAGE STANDARD DEVIATION, $\bar{s}$

Sample Size (No. of Observations per Sample) n	Factors for Control Charts for Averages			Factors for Control Charts for Standard Deviation		Factors for Process Standard Deviation C_4
	Based on Spec. Limits, F_{sd}	Based on Center Line		Upper Control Limit B_4	Lower Control Limit B_3	
		A_3	$\frac{2}{3}A_3$			
2	2.12	2.659	1.773	3.267	0	0.7979
3	2.20	1.954	1.303	2.578	0	0.8862
4	2.28	1.628	0.868	2.266	0	0.9213
5	2.35	1.427	0.951	2.089	0	0.9400
6	2.40	1.287	0.858	1.970	0.030	0.9515
7	2.44	1.182	0.788	1.882	0.118	0.9594
8	2.48	1.099	0.733	1.815	0.185	0.9650
9	2.51	1.032	0.688	1.761	0.239	0.9693
10	2.54	0.975	0.650	1.716	0.284	0.9727

11	2.56	0.927	0.618	1.679	0.321	0.9754
12	2.58	0.886	0.591	1.646	0.354	0.9776
13	2.60	0.850	0.567	1.618	0.382	0.9794
14	2.62	0.817	0.545	1.594	0.406	0.9810
15	2.63	0.789	0.526	1.572	0.428	0.9823
16	2.64	0.763	0.509	1.552	0.448	0.9835
17	2.66	0.739	0.493	1.534	0.466	0.9845
18	2.67	0.718	0.479	1.518	0.482	0.9854
19	2.68	0.698	0.465	1.503	0.497	0.9862
20	2.69	0.680	0.453	1.490	0.510	0.9869
21	2.70	0.663	0.442	1.477	0.523	0.9876
22	2.71	0.647	0.431	1.466	0.534	0.9882
23	2.71	0.633	0.422	1.455	0.545	0.9887
24	2.72	0.619	0.413	1.445	0.555	0.9892
25	2.73	0.606	0.404	1.435	0.565	0.9896
32	2.77	0.535	0.357	1.384	0.616	0.9919
50	2.83	0.427	0.285	1.322	0.678	0.9943

APPENDIX 4

CASES AND PROBLEMS

1. Lot-by-Lot Inspection of Aluminum Sheet Circles

In the blanking and annealing of aluminum sheet circles, problems were encountered as regards blisters, slivers, and scratches. Depending on magnitude, these deficiencies of a blanked disc would cause either a minor or a major defective. Acceptable quality levels (AQL) were 2 percent for major and 4 percent for minor defectives. Lot sizes were 400 sheet circles.

Required:
1. Determine and show sampling plans for single and sequential sampling.
2. Assume that sequential sampling is done. Inspection of an incoming lot reveals 1 major defective in the first sample of 40 and no defectives in the second sample of 10. What action should be taken with regard to major defectives?
3. Another lot has no major defectives, but the following minor defectives are found:

Sample	Size	Defectives
First	40	2
Second	10	1
Third	10	0
Fourth	10	1
Fifth	10	1

What action should be taken?

2. Control Chart for Sample Averages When Specifications are Stringent

Proper functioning of telephone ringers depends, among other factors, on the manufacturer's ability to keep the electrical resistance of ringer coils within close limits. For a particular type of coil, the tolerance was 3,400 ±300 ohms.

The higher the resistance, the greater is the amount of copper wire used in producing the coil. The manufacturer therefore had an interest in minimizing the amount of costly copper used, while at the same time conforming to specification limits.

Production is on two two-spindle machines. Each spindle has its own winding mechanism. During the week, six coils are selected randomly from each spindle, for resistance testing. For a five-week period, the results appear in Table A-1.

Table A4-1. Resistance, in Ohms, of Randomly Sampled Ringer Coils
(Data in 3,000 plus the Ohms Shown in Columns Below)

		Machine A			Machine B	
Week No.	Coil No.	Right Spindle	Left Spindle	Coil No.	Right Spindle	Left Spindle
One	1	440	380	1	460	390
	2	420	410	2	410	390
	3	480	390	3	430	380
	4	390	420	4	450	370
	5	410	370	5	420	360
	6	420	350	6	450	370
Two	1	520	390	1	380	360
	2	500	400	2	390	350
	3	490	380	3	410	290
	4	480	420	4	390	350
	5	540	380	5	380	290
	6	530	390	6	390	280
Three	1	480	500	1	490	390
	2	550	480	2	590	280
	3	560	460	3	480	280
	4	520	510	4	600	310
	5	550	490	5	570	290
	6	590	520	6	610	370
Four	1	710	590	1	210	090
	2	690	490	2	180	080
	3	720	400	3	190	110
	4	680	480	4	220	090
	5	810	390	5	190	090
	6	790	470	6	210	100
Five	1	580	580	1	280	290
	2	610	630	2	310	220
	3	580	590	3	310	270
	4	630	570	4	320	190
	5	600	620	5	330	190
	6	620	610	6	270	280

NOTE: An alternative analysis of this case is possible if the data above are assumed as coming from just one machine (say "A"). The entries under "Machine B" are then viewed as coming from Machine A, but for weeks 6, 7, 8, 9, and 10 successively.

We can now analyze the data by means of a control chart for sample averages, using the weekly machine totals and ranges. It makes more sense, however, to analyze the spindles separately. This will involve for each of the two machines:

1. Control chart for sample averages, for right spindle;
2. Same chart, this time for left spindle.

From the analysis, recommendations for improvement may be possible. These results would be concerned with both the average level of the process and the control chart limits for use in future testing.

You are asked to make the pertinent analyses and submit appropriate recommendations based on your findings.

3. Control Chart for Sample Averages and Ranges

In the manufacture of low voltage electrical fuses, an important quality characteristic is the average and range of blow time, in seconds, under specified overloads. Pertinent data regarding a particular type of fuse are given in Table A-2. Over a period of 30 days, four sample fuses were selected randomly and tested.

Table A4-2. Low-Voltage Electrical Fuse Quality

Day No.	Blow Time, in Seconds — Four Randomly Selected Fuses				Day No.	Blow Time, in Seconds — Four Randomly Selected Fuses			
1	35.4	35.3	34.6	35.5	16	34.2	34.3	34.4	34.1
2	35.0	35.0	34.8	35.0	17	34.9	34.7	34.4	34.8
3	34.3	34.3	34.2	34.0	18	33.5	34.1	34.0	34.4
4	34.2	34.6	34.8	34.4	19	34.4	34.4	33.8	32.8
5	34.4	34.7	34.4	34.3	20	34.7	34.1	34.4	34.2
6	34.5	34.2	34.7	34.4	21	33.4	34.4	33.8	33.6
7	34.4	34.8	34.5	34.5	22	32.9	33.8	33.6	33.3
8	34.6	34.8	34.4	35.8	23	34.4	34.1	34.7	34.2
9	34.8	34.5	34.8	35.1	24	34.2	33.4	33.3	33.9
10	34.7	34.9	34.7	35.3	25	33.0	33.3	33.7	33.2
11	35.3	34.9	34.5	34.7	26	34.2	33.9	34.1	34.2
12	34.2	34.1	34.0	34.1	27	34.7	34.1	34.7	33.3
13	34.3	35.4	34.4	34.7	28	34.9	35.8	35.1	34.8
14	34.9	34.3	34.3	34.5	29	34.0	33.9	34.1	34.6
15	34.3	34.6	34.1	34.2	30	34.7	34.1	33.9	34.3

In order to analyze performance, proceed as follows:

1. Use the first ten days' data to obtain the grand average of the process, the average range, and upper and lower control limits for sample averages and ranges.
2. Plot the first ten days' data to note whether or not the process is in control.

3. Extend the control limits for 20 additional days (to the total of 30 days).
4. Note all out-of-control conditions.

Since the fuse is of a relatively novel type, no specification limits have been firmly established as yet. Nevertheless, from the 30 days' experience, recommendations are possible as to the degree of control over process average and range that may be attainable.

Make the appropriate recommendations.

4. Process Capability of Coil Winding Operation

In a prior case (see page 373), we obtained test results of resistance, in ohms, of the ringer coils produced on two two-spindle winders. Additional worthwhile information can be won from the tabulated values for the purpose of appraising process capability. For this purpose, one should proceed as follows for each of the two machines:

1. Prepare frequency distribution graphs for (A) right and (B) left spindle (separately).
2. Calculate the standard deviation and variation coefficient of each distribution.

From the results, the variability maintainable by the process under normal operating conditions can be ascertained.

In plotting the frequency distributions, class intervals of 3,000 ohms plus the ohms shown below are recommended: 000 to 099, 100 to 199, 200 to 299, ... etc. ... 3,800 to 3,899.

Required: Make recommendations to management regarding the following:

1. Average level toward which processing at each spindle should aim.
2. Tolerance that can be maintained.

The recommendations should be consistent with the tolerance requirements of 3,400 ± 300 ohms, while at the same time minimizing unnecessary consumption of costly copper wire.

5. Control Chart for Percent Defective

A manufacturer maintains 100 percent inspection of completed electromechanical products. In a typical day, 1,000 units are inspected. From past experience, it has been found that the average proportion of parts requiring replacement is 2 percent. Excessive replacements may indicate one of the following undesirable conditions:

1. Production of defective parts exceeds allowable levels.
2. In-process inspection is not efficiently screening out defectives.
3. Incoming purchased items are not of acceptable quality, and receiving inspection is not successful in keeping these defectives out of the plant.
4. Instead of making the necessary adjustments on the completed products, the final inspection (which includes adjustments and replacements) relies on replacements when these are not really needed.

Required: Prepare a control chart with 3-sigma limits, which management may use in daily surveillance of quality and operating efficiency. What is the value, in percent, of the upper control limit?

6. Control of Defects per Unit in an Electronic Shop

In the production of electronic modules, the potting process was causing special problems. Over a number of weeks, the following was recorded:

Week No.	Units Produced	Defects Observed Broken Welds	Dimensions Off	Holes & Chips	Other	Total
1	150	5	1	4	10	20
2	200	8	0	0	6	14
3	180	9	3	6	6	24
4	225	10	4	2	20	36
5	200	4	1	0	5	10
6	300	2	10	0	2	14
7	230	12	2	2	4	20
8	150	15	5	5	0	25
9	175	5	3	0	3	11
10	190	0	1	1	4	6
Total	2,000	70	30	20	60	180

As an aid for management, develop control chart limits for (A) total defects per week and (B) defects per week for each of the four categories observed.

7. Variance Analysis of Coil Winding Operation

For the purpose of improving the quality and efficiency of ringer coils produced on a two-spindle machine, we developed control charts (page 373) and studied the process capability (page 375). It is apparent, however, that an analysis of variance will provide further worthwhile information of practical usefulness to production management, engineering, and supervision.

Required: Perform an analysis of variance for each of the two winding machines, with particular attention to the following:

1. Within-spindle and machine-overall variation, as reflected by average range, standard deviation, and coefficient of variation.
2. Between-spindle variation, obtained by subtracting the overall variance from the within-spindle variance and expressing the result in standard deviation and variation coefficient.
3. Evaluate the significance of the excess of the overall versus the within-spindle variation.
4. Summarize the salient results of the analysis. In what manner do they supplement prior information? How will this information help production management, engineering, and supervision?

8. Variance Analysis of Castings

Extruded lengths of an alloy were tested for tensile strength. For three weeks, two casts were tested per week, and the strengths in 1,000s of pounds for three lengths were recorded:

Week No. 1		Week No. 2		Week No. 3	
Cast No. 5	Cast No. 1	Cast No. 3	Cast No. 7	Cast No. 8	Cast No. 4
37	38	38	36	39	36
38	38	36	37	38	38
39	36	37	38	38	37

Production management was concerned about the possibility of high between-cast variation as a result of any one of the following factors: (1) inadequate stirring of furnace load, (2) variations in temperature of cooling water, (3) fluctuations in flow rates and (4) nonuniformity of other casting conditions.

It is therefore desirable to evaluate within-cast, between-cast, and overall variation.

Required:
1. Find the standard deviation and variation coefficient for within-cast, between-cast and overall variation.
2. Develop control charts that supplement the analysis of variance results.

9. Tolerance Chain for Pulley and Gear Assembly

A pulley and gear assembly on a shaft consists of the following arrangement: spacer, gear, spacer, pulley, spacer. The spacer can be produced readily to a process standard deviation of 0.0005 inch, at a cost of $0.50.

For the gear and pulley, however, different processes at different costs apply:

	Gear	Pulley
Standard deviation, inch, at a cost of $2.00 per gear	0.002	...
Standard deviation, inch, at a cost of $3.00 per gear	0.001	...
Standard deviation, inch, at a cost of $1.50 per pulley	...	0.005
Standard deviation, inch, at a cost of $2.00 per pulley	...	0.003
Standard deviation, inch, at a cost of $2.50 per pulley	...	0.002

Required:
1. Tolerance, overall, in inch, assuming that cost is unimportant.
2. Tolerance, overall, that minimizes cost.
3. Minimum cost tolerance combination, with the provision that overall tolerance must not exceed ±0.010. The tolerance values for the gear and the pulley should be shown separately, followed by the overall tolerance (including the spacers) and the total cost.

10. Tolerances in Piston Production

A piston is to be produced to an overall dimension of 3 inches. The following are further pertinent data:
1. Overall dimension must be maintained to a tolerance of 0.005 inch.
2. Next, a pinhole must be fine-bored on a separate machine. The location of the hole is specified to be 1 inch from the piston top. A tolerance of 0.003 inch can be maintained on this operation.

The important final dimension is the distance of the pinhole from the top of the piston.

What is the tolerance that can be maintained for the final dimension of the pin distance?

11. Tolerancing for Electronic Assemblies

Assembly of an electronic device includes the combination of an attenuator (step-down of 0.2) and a transformer (amplification of 10). Input voltage is 100, while output voltage is 200, based on the relationship:

Output = Input × Attenuator × Transformer
$$= 100 \times 0.2 \times 10$$
$$= 200 \text{ volts}$$

A circuit diagram appears in Fig. A-1.

Parts tolerances are as follows: Input, ±2 volts, attenuator, ±0.01 volts and transformer, ±0.1 volts.

It is now necessary to evaluate the variation in output voltage to be expected for the assembly. This variation should be expressed as the

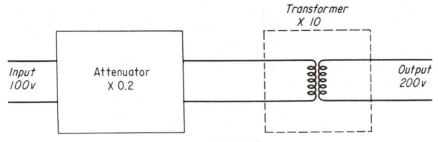

Fig. A 4-1.   A circuit diagram.

anticipated tolerance that the assembly can maintain during normal operations.

Required: Perform the necessary evaluation of overall tolerance for the assembly, using two approaches:

1. Analytical solution, using the formulas for tolerance combination given in the text.
2. Simulation solution, using the type of approach shown on page 211 of the text. (Instead of a computer, manual combination of simulated outputs for about 100 sets of input, attenuator, and transformer tolerances is possible quite readily.)

12. Evolutionary Process Operation

Magnetic output, in millivolts, of a core used in memory circuits was to be increased without major changes in processing. An evolutionary operations phase yielded the results below:

Setup	1	2	3	4	5
Time, seconds	7	6	8	8	6
Temperature, °C	400	300	500	300	500
Cycle		(Output, mv.)			
1	12	12	15	13	9
2	10	8	13	14	11
3	12	10	14	11	10
4	12	11	13	11	8
5	10	9	13	10	12
6	10	10	16	13	10
Total, mv.	66	60	84	72	60
Average, mv.	11	10	14	12	10

Setup (3) yields the highest average output in millivolts. The question, however, is whether this is a significant improvement. In order to answer this point, we must determine:

1. The error limit, in millivolts.
2. The effect of time, in seconds, on millivolt output.
3. The effect of temperature, in degrees centigrade, on millivolt output.
4. The magnitude of the effect in (2) and in (3) in comparison to the error limit.

Only if the effect of time or of temperature exceeds the error limit, is the effect significant. What conclusions should be drawn?

13. Evaluation of Wear-Out Failure

Length of life of an electrical, stranded wire depended primarily on wear-out from repeated flexures. Starting with 50 wires, each of which lasted the absolute minimum of 1,000 of flexures, the following failures were observed:

Testing Interval Observed (Number of Flexures Applied to the Wire Strand)		Number of Failures (Breaks) Occurring During Each Interval
Start	End	
1,000	1,999	1
2,000	2,999	3
3,000	3,999	8
4,000	4,999	20
5,000	5,999	10
6,000	6,999	6
7,000	8,000	2

In order to evaluate life quality of the wire, the following is now required:

1. Cumulative failures and survivals, actuals and percent.
2. Probability (based on data obtained from step 1) that a wire will withstand the following flexures or more without failing: 1,500, 2,500, 3,500, 4,500. Probability of failure before 1,600 flexures.
3. Force of failure, percent.
4. Standard deviation and variation coefficient of wire life.
5. Reevaluation of probabilities (step 2 above), assuming wire life and wear-out are normally distributed.
6. A plotting on probability paper (try normal and Weibull grids).

14. Reliability Evaluation for Constant Failure Rates

For a certain type of potentiometers produced by a manufacturer, a constant failure rate has been observed during the useful life of the device. Accordingly, the failure distribution has a negative exponential form. During its life, the potentiometer has a 2-percent chance of failure. A total of 100 potentiometers are placed in service for a 1,000-hour time span. It is now necessary to evaluate the following:

1. Number of potentiometers that will fail during a period of 1,000 hours.
2. Number of failures to be expected during the total of 1,000 times 100 or 100,000 hours of operation.
3. The number of failures per service hour, which is the failure rate, λ (lambda).
4. Mean time to failure, MTTF or m, representing the average expected time interval between failures.
5. Accuracy of calculations, by checking that m times λ does indeed equal 1.0.
6. Ratio t/m, assuming that the required operating time of the potentiometer is 50,000 hours.
7. Probability of survival of the potentiometer for the following number of hours: (A) 5,000, (B) 2,000, (C) 1,000, and (D) 500 hours.

Both the reliability curve with ratio t/m at the base and the table of reliabilities in Chapter 16 should be used, where applicable.

15. Testing for Reliability

Reliability of a shipment of chip-size integrated circuits is to be assured by means of testing. A mean life of 10,000 hours per unit has been specified. Slightly over three weeks are available for testing, corresponding to approximately 500 hours. The failure distribution may safely be assumed to be exponential. A 90-percent confidence level, or in other words a 10-percent risk of erroneously rejecting a good lot, is considered appropriate.

The following should now be determined:

1. The ratio of testing time to mean life.
2. Sample size (number of units to be tested), assuming we wish to reject the shipment if the number of failures exceeds one.
3. Sample size if we wish to reject the shipment on the basis of at least the following number of failures: 2, 3, 4, 5, 10.

What are the practical limitations to reliability assurance by means of the types of sampling and testing just discussed?

16. Evaluation of System Reliability

The probability of success, P_s, of a system hinges on the component reliabilities and the manner in which they are arranged, such as series, parallel, or series-parallel combinations. We will consider typical problems below.

Problem 1: Series. Three parts A, B, and C have the reliabilities $P_s(A) = 0.7$, $P_s(B) = 0.8$, and $P_s(C) = 0.9$. We read $P_s(A)$ as "the probability of success of component A," where "success" refers to proper functioning under specified conditions for a certain period of time. The components are arranged in series, so that system reliability is simply the product,

$$P_s(\text{System}) = P_s(A) \times P_s(B) \times P_s(C)$$

Find the value sought.

Problem 2: Parallel. When calculating reliability of components arranged in parallel, we must also consider the probability of failure, P_f. For example, $P_f(A) = 1 - P_s(A)$ which, for our example, is $1 - 0.7$ or 0.3. Now, if A and B are in parallel, then $P_f(A \text{ and } B) = P_f(A)$ times $P_f(B) = 0.3 \times 0.2 = 0.06$. Next,

$$P_s(\text{System}) = 1 - P_f(\text{System}) = 1 - P_f(A) \times P_f(B)$$

which, for our example, is $1 - 0.06$ or 0.94. Now calculate the reliability of a system in which A, B, and C are in parallel.

Problem 3: Simple Combination. Consider a system in which A and B are in parallel, followed by C and D in series, with $P_s(D) = 0.6$. For this purpose, we may use the value 0.94, found above, for $P_s(A \text{ and } B)$ and consider it part of the series, such that

$$P_s(\text{System}) = P_s(A \text{ and } B) \times P_s(C) \times P_s(D)$$

Apply the formula to assess system reliability.

Problem 4: Compound System. In the diagram of Fig. A-2, the two subsystems A-B-C and E-F are each in parallel.

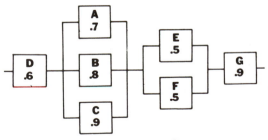

Fig. A 4-2. A system diagram.

In order to appraise system reliability, we must first assess the reliabilities of *A-B-C* and *E-F*. Note that for *A-B-C* we already found the requisite value (from problem 2). Next, we use the subsystem reliabilities as part of a series. Thus:

$$P_s(\text{System}) = P_s(D) \times P_s(\text{Subsystem } A\text{-}B\text{-}C) \times$$
$$P_s(\text{Subsystem } E\text{-}F) \times P_s(G)$$

You are asked to substitute the appropriate values and calculate the system reliability.

Problem 5: Compound System, Further Example. For the system portrayed in Fig. A-3, appraise overall reliability.

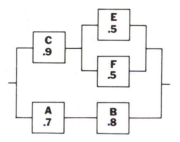

Fig. A 4-3.   A compound system.

Suggestion: You already know $P_s(E \text{ and } F)$ from the preceding problem. Next, consider the series *C* and *E-F* as a subsystem, similar to the subsystem *A* and *B*. You are left with a simple system in parallel, consisting of *one* P_s value for *C* and *E-F* and *one* P_s value for *A* and *B*. Nothing more than the method of problem 2 will now be needed to obtain overall reliability.

17. Three-Factor Replicated Experiment

A tire manufacturer, interested in enhancing the bruise energy characteristics of passenger tires, investigated the factors of (a) size of yarns in terms of deniers, (b) number of yarns per inch, expressed as "ends per inch," and (c) cured cord angle in degrees. Bruise energy is measured in 100s of inch pounds. The results of a replicated experiment appear below:

Ends Per Inch	Angle, Degrees	Denier 840/2				Denier 1000/2				Angle Mean
		Run 1	Run 2	Range	Mean	Run 1	Run 2	Range	Mean	
24	31	20	22	2	21	24	22	2	23	22
24	33	21	19	2	20	22	22	0	22	21
24	35	17	19	2	18	18	20	2	19	18.5
24	37	10	16	6	13	16	16	0	16	14.5
24 - Ends/Inch Mean					18				20	
26	31	24	22	2	23	24	24	0	24	23.5
26	33	23	19	4	21	22	24	2	23	22
26	35	19	19	0	19	17	23	6	20	19.5
26	37	18	20	2	19	21	17	4	19	19
26 - Ends/Inch Mean					20.5				21.5	
Denier Mean					19.5				20.25	

	Further Means			Grand Mean
Angle Degrees	Mean for 24 Ends/Inch	Mean for 26 Ends/Inch	Combined Angle Means	
31	22	23.5	22.75	
33	21	22	21.5	
35	18.5	19.5	19	
37	14.5	19	16.75	
Combined Ends/Inch Means	19	21		20

For management to properly appreciate these data, it is now necessary to

1. Estimate the experiment error standard deviation and decision lines.
2. Plot the main effects as well as the two-way and three-way interactions.
3. Draw in the decision lines (at 95 percent confidence level).
4. Briefly report the findings from this analysis.
5. Submit recommendations for appropriate choice of a standard tire design. For this purpose one must recognize that a denier of 1000/2 is more costly than a denier of 840/2, that 26 ends per inch cost more than 24 ends per inch, and that there is no cost differential among the four cord angles.

You have been asked to make the analyses and evaluations needed, as shown above, and to report your findings and recommendations.

Glossary

Accelerated Test. A test in which the stress level applied exceeds that needed in practical use, in order to shorten the time required to observe the stress response of the product. A valid accelerated test must not alter the basic mechanisms of failure or their relative prevalence.

Acceleration Factor. The ratio between two different sets of stress conditions involving the same failure mechanisms.

Acceptable Quality Level, AQL. The level at which a lot contains up to the proportion of defectives considered acceptable.

Acceptance. Decision that a lot is of acceptable quality. A bad lot may be accepted as a result of sampling error.

Acceptance Control Chart. *Control chart* used in acceptance inspection, often permitting a (usually small) proportion of defectives as an *acceptable quality level, AQL.*

Acceptance Number. A number or other value against which a sampling outcome is checked. In general, if the defectives in a sample are equal to or less than the acceptance number, the lot is considered acceptable.

Allowance. Difference between mating part dimensions, permitting a specified maximum amount of looseness or tightness.

Alpha, α. Producer's risk of erroneous rejection of a good lot.

Alternate Hypothesis, H_a. Hypothesis that there is a relationship between two or more factors. This H_a is established by rejecting the *null hypothesis.*

Arithmetic Mean. A measure of central tendency of a group of units, obtained by summing the units and dividing the total by the number of units involved.

Assignable Cause. A cause that can be found, in the sense that a particular source of faulty output may be identified.

Attributes Method. Measurement of quality of a unit of product by noting whether or not a specified characteristic or "attribute" is present or absent. A unit may, for example, be "effective" or "defective." Go/no-go testing is in effect attributes testing.

Availability. Ratio defined as $m/(m + r)$ where m is mean time between failures and r is the mean downtime for repair.

Average Outgoing Quality Limit, AOQL. Term applicable in lot-by-lot sampling when rejected lots are screened. When the theoretically perfect lots (from screening) are combined with the lots containing variable proportions defectives (passed by sampling), the long-run quality leaving the inspection station is the AOQL, in percent of defectives.

Average Sample Number, ASN. Average sample size expected in a sequential sampling plan.

Basic Dimension. A size to which a tolerance is applied.

Bayesian Probability. A rule for finding *posterior probabilities*, as illustrated below: *Prior probabilities* that a lot is of acceptable or unacceptable quality are $P(A) = 0.8$ and $P(U) = 0.2$. Corresponding *conditional probabilities* of finding 2 defectives in the sample, given that the lot is A or U, are $P(2|A) = 0.1$ and $P(2|U) = 0.5$, yielding *joint probabilities* $P(2 \text{ and } A) = 0.8 \times 0.1 = 0.08$ and $P(2 \text{ and } U) = 0.2 \times 0.5 = 0.1$. The sum of the joint probabilities is $P(2) = 0.18$. Bayes rule now says that $P(A|2) = P(2 \text{ and } A)/P(A) = 0.08/0.18 = 4/9$. Further, $P(U|2) = 0.1/0.18 = 5/9$. Finally, the sum of the posterior probabilities $4/9 + 5/9$ checks out to 1.0, as required.

Beta, ß. Consumer's risk of erroneous acceptance of a bad lot.

Beta Distribution. A probability distribution.

Binomial Distribution. A probability distribution that applies when random samples are identified in terms of two classifications, such as "effective" and "defective."

Catastrophic Failure. Sudden change in operating characteristics resulting in complete loss of useful performance of the function or unit.

Censored Data. Data representing incomplete life tests. During testing, some units will have failed, and we can determine the time of failure. For those units, however, which are still operational, the failure point is some time in the future. Censored data are thus incomplete life testing results.

Center Line. Line on a control chart representing the grand average or standard or goal value for the statistical measures being plotted.

Chance Failure. After all efforts have been made to eliminate design deficiencies and remove unsound components, and before wear-out becomes a factor, there may yet be occasional failures. Usually, these will occur at random intervals and are therefore called "chance."

Clearance. Situation in which the amount of metal, such as in a shaft, is less than in the dimension to which it is to be assembled, such as a hole.

Conditional Probability. The likelihood that a particular cause has been responsible for an event. Conditional probabilities give us this likelihood. For example, in tossing a fair, six-sided die, what is the probability that the number obtained is divisible by 3, given that the number is even. The probability of an even number, $P(E)$, is obtained by noting that among the six mutually exclusive outcomes, 1, 2, 3, 4, 5, 6, there are 3/6 even numbers. Hence, $P(E) = 1/2$. The probability of a number divisible by three, $P(D)$, is obtained by noting that only 2 of the 6 numbers have this characteristic, so that $P(D) = 1/3$. Next, the *joint probability* $P(D$ and $E) = (1/2)(1/3) = 1/6$. Finally, the *conditional probability* of $P(D|E) = $ (Joint Probability)$/P(E) = (1/6)/(1/2) = 2/6 = 1/3$.

Confidence Level. Probability that a particular value lies between an upper and a lower bound, the *confidence limits.*

Confidence Limit. The bounds of an interval. A probability can be given for the likelihood that the true value will fall within the interval.

Control Chart. Graphic record on which we plot successive *statistical measures* or other values of a *random variable.*

Control Limit. Limit on a *control chart* for judging whether or not a *statistical measure,* obtained from the *sample,* falls within acceptable bounds.

Critical Defect. Defect that could result in hazardous or unsafe conditions.

Defect. An instance of failure to satisfy a quality specification.

Defective (Unit). A unit that contains one or more defects.

Degradation Factor. Factor by which actual system failure rate differs from inherent reliability. For example, the bench test reliability of a piece of equipment is measured repeatedly and found to have a mean life of 40 hours. In actual use, however, the same equipment demonstrates a mean time between failures of only 20 hours. Degradation factor is therefore $40/20 = 2.0$.

Degrees of Freedom, DF. Term referring to the number of observations in a sample that are not fixed. For example, when the variance of a population is estimated, DF equals $n - 1$, where n is the sample size.

Derating Factor. Factor by which equipment is derated to achieve a comfortable reliability margin of safety. For example, from a pure design viewpoint, a probability of failure of 0.05 may be acceptable. In order to achieve a 2 to 1 safety margin, however, a probability of failure of 0.025 would have to be asked for. The derating factor is therefore 2.

As a further example, if a 2-watt resistor is used when a 1-watt unit is adequate, the derating factor is 2.

Double Sampling. A form of sequential sampling in which two samples are taken from a lot.

Dynamic Programming. Investigation of problem variables involving several successive phases or levels of operation. An optimum combination of factors, considering all stages, is sought.

Effectiveness. The product of reliability times availability for a given level of performance under specified operating conditions.

Evolutionary Operation. Method of experimentation during production, whereby process variables are changed in a systematic manner, and the results are observed. The changes are small, in order to avoid the risk of off-standard product or other undesirable effects. Statistical analysis provides guides toward process improvement in the form of increased yield or improved quality.

Exponential Distribution. Probabilitiy distribution for which the force of failure or *hazard rate* is constant.

Exponential Function. A function with an exponent. However, when we talk about *the exponential function,* we refer to a function that has e, the base of the natural logarithm, as the quantity that is raised to a given power. Some interesting aspects about e are:

1. $e = $ limit of $(1 + 1/n)^n$ as n approaches infinity.
2. The exponential function is the only function that is exactly its own derivative. In particular, the derivative of an exponential function, $f(x)$ is given by the product of that function and some number, say, L, such that $f'(x) = f(x)(L)$, where $f(x) = a^x$. Now, that special value of a that causes L to be unity is e. Thus, if $f(x) = e^x$, then $f'(x) = f(x)$. It can be shown that e does in fact exist and is an irrational number (that is, it cannot be written as the ratio of two integers, also known as a "never-ending decimal").
3. From these definitions, it is apparent that the exponential function describes all such real-world problems where a quantity changes at a rate that is proportional to the quantity itself.
4. Further, e raised to the power $(i\,\pi)$ equals -1 where $i = \sqrt{-1}$. Yet though e has been computed to 60,000 decimals, no orderly pattern has been found.

Exponential Smoothing. Method for smoothing out the fluctuations in a data series, such as failure rates over time. The process is akin to "averaging out" extremes of fluctuations.

Failure. The termination of the ability of a product to perform as required, because of misuse, inherent weakness, sudden factors that could not be anticipated, or gradually developing factors. Failure may be partial or complete; it may be catastrophic (sudden and complete), or it may represent degradation (partial and gradual).

Failure-Mode Analysis. For each critical parameter of a system, the determination of which malfunction symptoms appear just before or immediately after failure. For each symptom, the designer lists all possible causes and then attempts to design the product to eliminate the problem. Failure-mode analysis is of an anticipatory nature, occurring before actual production in the design stage.

Failures, Percent per 1,000 Hours. The number of failures per 1,000 hours multiplied by 100, or in other words the number of failures per 100,000 hours. This is a peculiar wording that has gained wide usage. In *no way* does it refer to the percent of parts in a device.

Failure Probability. Probability of failure in a given time period.

Failure Rate, λ. The number of times a piece of equipment is likely to fail in a given time period. In order to ascertain failure rate, a sample piece is run for a certain time, such as 1,000 hours, and the number of failures occurring in that period is recorded. After each failure, the equipment is repaired. If an equipment is not repairable, then the failure rate can be ascertained by testing several pieces of equipment and recording the total time of run for all pieces until failure.

Fraction Defective. Proportion of defective units, expressed as a decimal. If a sample contains 10 percent defectives, the fraction defective is 0.10.

Frequency. Number of times an event occurs. Assume, for example, that a die is thrown 36 times. On six occasions the number 2 occurs. The frequency of the event is thus 6. The relative frequency is $6/36 = 1/6 = 0.1667 = 16.67$ percent.

Frequency Distribution. Representation of a universe or sample, whereby individual values are grouped into equal-sized classes. The number of values falling within each class interval are recorded.

Go/No-Go Testing. Method of test, involving an inspection gage or instrument with limits to determine whether or not a measurement (1) conforms or (2) fails to conform with specifications.

Hazard Rate. Rate that is identical with force of failure for the exponential distribution (for others, this is true only approximately). The exact term depends on an understanding of the symbol d in connection with dt as an infinitesimal increase in t, where t is usually a time interval. For a similar increment, dN_f, for the number of failures in a life test, the hazard rate becomes dN_f/dt. Another term used for hazard rate is *instantaneous failure rate*. When such rates are constant, as in the case of the negative exponential distribution, the word "instantaneous" is often omitted.

Infant Mortality. Early failures, until debugging and shake-out have eliminated faulty components, improper assemblies, and other such problems; and a more normal rate of failure prevails.

Inherent Reliability. Reliability potential of a design.

Inspection Level. Term indicating the sample size required. A high inspection level requires a large sample size and thus lower risk of erroneous acceptance of off-standard product.

Inspection Lot. Collection of units, made under substantially the same conditions, ready for inspection.

Instantaneous Failure Rate. Same as *hazard rate*.

Interference Fit. Situation in which the amount of metal, such as in a shaft, exceeds the amount allowable for the fit, such as in a hole.

Joint Probability. Probability that two events occur together. Given independent events, the joint probability is the product of the individual probabilities. For example, a system consists of two independent functions A and B. Failure probabilities are 0.2 and 0.4 respectively. The joint probability that *both* will fail is 0.2 $\times$ 0.4, or 0.08.

Lambda λ. See *failure rate*.

Limit Design. Provision to allow for the "worst case" of stresses that might occur, considering safety-factor allowances.

Lot-by-Lot Inspection. Successive output is grouped in the form of lots of approximately equal number of *units* each, and quality is evaluated by means of sampling.

Lot Size. The number of individual units in a lot.

Lot Tolerance Percent Defective, LTPD. Somewhat of an anachronism, this term defines an unacceptable quality level, at which it is desired that the probability of rejection of the lot be high (usually, 90 percent or better).

Maintenance Capability. The facilities and trained personnel, as well as the engineering support and spare parts availability, to keep a system in serviceable condition.

Maintenance Ratio. The number of maintenance man hours of downtime required per hour of regular operation of a system. The failure rate of the system, together with the amount of time required to analyze and repair breakdowns, is reflected. When the ratio exceeds 1, the system is down more than it is "on."

Malfunction Analysis. Study of a failure after it has occurred, with a view to developing correctives in design of future devices.

Mean Time Between Failures, MTBF or M or μ. The average time interval between failures for repairable equipment, usually measured in terms of operating hours. For a large number of devices, particularly in the electronic field, failure time probabilities are distributed according to the negative exponential distribution. Note that MTBF is the reciprocal of the failure rate. For example, if an equipment has a failure rate of 10 per 1,000 hours, or 1 per 100 hours, then the mean time between failures is the reciprocal of $1/100$ hour $= 100$ hour. Had the failure rate been 5 percent per 1,000 hours, we would have found: Failure rate is 5 per 100,000 hours $= 5/100,000$ hour. Next, the reciprocal of $5/100,000$ hour is $100,000$ hour$/5 = 20,000$ hour. The MTBF is particularly useful in evaluating end use conditions, by permitting an estimate of the time interval in which equipment is likely to fail, so that service and preventive maintenance (replacement of tubes, other parts) can be planned accordingly. As a corollary, the number of spare and replacement parts needed can also be determined.

Mean Time to Failure, MTTF or MTF. The average time until a non-repairable device fails, measured in operating hours. Again, MTTF is the reciprocal of the failure rate. While most equipment is either repairable or nonrepairable, some items can have both states. For example, on the ground a spacecraft is repairable; once aloft it is non-repairable. Moreover, a missile or spacecraft in flight will encounter more environmental stresses (shock, vibration, temperature, radiation) than on the ground. As a result, higher stresses on equipment may mean higher failure rates, thus less MTTF. Nonspace examples of MTTF consideration are many: MTTF of a plastic heart valve or an inserted electronic pacemaker, life of a bridge or other construction, reliability of a tire or battery.

Mean Time to Restore MTR. Average time interval to repair equipment and restore it to regular operation.

Median. The middle value in a set of quantities arrayed from smallest to largest. Assume for example, the following temperature readings in centigrade: 71, 78, 68, 80, 65. Arrayed, these are 65, 68, 71, 78, 80. The middle value of 71 is the median.

Mode. That value in a *frequency distribution* which represents the most frequently occurring class.

Multiple Sampling. Another term for *sequential sampling.*

Natural Environment. External conditions, such as temperature, humidity, pressure, radiation, wind, and the like.

Negative Exponential Distribution. See *exponential distribution.*

Normal Curve. Bell-shaped curve, representing a frequency distribution developed from theoretical considerations (the so-called "normal law of error"). Its significance stems from the fact that, in approximate form, the distribution is found to occur widely in natural and manufacturing phenomena.

Null Hypothesis, H_o. An initial assumption in the statistical testing of significance, whereby it is held that there is no relationship between two or more factors.

On-Line Maintenance. Maintenance work performed during operation of the equipment.

Operating Characteristic. The manner in which a sampling plan will produce acceptances or rejections of lot, corresponding to the quality of the lot inspected.

Percent Defective. Proportion of defective units in a lot.

Poisson Distribution. Probability distribution that often applies when the distribution of rates is involved. The Poisson distribution is the reciprocal of the negative *exponential distribution.*

Posterior Probability. Estimate of probability based on a combination of prior probability plus sampling outcomes.

Precision of Estimate. Magnitude of an interval within which the universe value, such as a lot mean or standard deviation, can be expected to fall at a given level of probability. Precision of estimation of averages increases with the square root of the number of observations.

Prior Probability. Estimate of probability prior to sampling based on objective or subjective information.

Probability. Likelihood of occurrence of a particular event. Probability is usually given as a ratio of (1) the number of ways an event actually occurs to (2) the total number of possibilities. For example, in tossing an unbiased die, a *two* can occur one way, but a total of six outcomes would be possible. Hence, the probability of a two is $1/6 = 0.1667$ or 16.7 percent.

Probability of Failure, P_f. Probability that equipment will fail. In general, $P_f = 1 - P_s$, where P_s is the *probability of success.*

Probability Paper. Paper with special grids, facilitating the plotting of probability distributions, such as beta, binomial, normal, or Weibull.

Probability of Survival, P_s. The likelihood that a particular device will survive for a given period of operating time. In all those instances where a negative exponential failure distribution may be expected, we may state that the probability of survival will be:

$$P_s = e^{-t/m}$$

where "t" is the operating time of the device, $m = $ MTBF or MTTF and $e = 2.71828$, the base of the natural logarithm. The curve above is also known as the "constant decay rate curve," (frequently used in operations research work, such as in analyzing "sales decay" of a product that is no longer being advertised). Since the reciprocal of e is 0.368, we may define P_s also as:

$$P_s = (0.368)^{t/m}$$

If the MTTF of a missile is 1,000 seconds, then the probability of its surviving 1,000 seconds is:

$$P_s = (0.368)^{1,000/1,000} = 0.368^1 = 0.368 = 36.8 \text{ percent}$$

The probability of its surviving 2,000 seconds is:

$$P_s = (0.368)^{2,000/1,000} = 0.368^2 = 13.5 \text{ percent}$$

The probability of its surviving 500 seconds is:

$$P_s = 0.368^{1/2} = \sqrt{0.368} = 0.607 \text{ or } 61 \text{ percent}$$

Observe an important result: When an equipment has a particular MTBF or MTTF, then the chances that the equipment will survive for that length of time, m, are not (as one might have expected) 50 percent but merely 36.8 percent.

Process Capability. Proportion of defectives or quality variation that may be expected from a production process working at the settings, speeds, materials inputs, and other operating arrangements specified.

Process Inspection. Periodic examination and measurement with emphasis on the checking of processing variables.

Quality. Quantitative levels or degrees of conformance and nonconformance of manufactured product, raw materials, services, or other items and operations.

Quality Assurance. The total effort of *quality control* and *quality engineering.* Often, however, the terms quality control and quality assurance are used interchangeably.

Quality Control. Activities designed to minimize the incidence of nonconformance during and after production. Specifications and tolerances are established, process capabilities ascertained, and tests and inspections performed to compare actual against standard performance. In instances of off-standard output, a response mechanism provides for (1) correction and (2) steps to prevent recurrence of defects. The term is often used interchangeably with *quality assurance*.

Quality Engineering. Analysis of the manufacturing system at all stages of development, including evaluation of designs, process optimization studies, and development of process capabilities. Quality engineering utilizes such analytical techniques as reliability evaluation, human factors studies, operations research, and cost and effectiveness analyses.

Random Variable. A number or other quantity whose outcome is dependent on a probability distribution.

Range. Difference between the largest and smallest value in a group.

Redundancy. Provision of duplicate, triplicate, or more elements to share a load, even though each element alone can do⁻ the work. When one element fails, the duplicate unit can take over.

Rejection. Decision that a particular lot cannot be considered as being of acceptable quality. A good lot may be rejected as a result of sampling error.

Rejection Number. Criterion for rejection of a lot. In general, when defectives or other failures in a sample equal or exceed the rejection number, the lot is rejected.

Reliability. The probability that a product will perform a required function under stated conditions for a given period of time. It may be assessed within stated confidence limits from tests or field failure experience, and it may be predicted from design considerations and from computations using component part reliabilities.

Reliability Index. Any factor that denotes reliability, such as *mean time between failures* or *the failure rate*.

Repairability. Probability that a failed system will be restored to operable condition within a specified period of active repair work.

Risk. Probability of error. Two types of risk are possible. A good lot may be rejected, or a bad lot may be accepted, as a result of sampling fluctuations or the "luck of the draw" in sampling.

Another definition of risk is that it represents a situation in which probabilities can be attached to a likely *state of nature* (see also *uncertainty*).

Sample. A group of *units* or portion of material, taken randomly from a *universe*. Units need not be physical. For example, a sample may consist of a number of measurements or other observations.

Sample Size. Number of *units* in a *sample*. When a sample of 10 units is taken from a population of 1,000 units, for example, we say that the sample size is 10.

Sampling Table. Tabulation showing *sample size, acceptance,* and *rejection numbers.*

Screening. The inspection of every unit of product in a lot.

Sequential Sampling. Sampling technique in which a succession of samples, of a particular size each, are chosen at random. See also *single* and *double sampling.*

Significance Testing. Statistical appraisal of the outcomes of sampling to note whether or not, at a certain level of risk, the results represent real effects or chance fluctuations of sampling and measurement. See also *hypothesis testing.*

Simulation. Replication in miniature form, usually on a computer, of a sizable real world problem. By investigating factors and variables at different levels, a comprehensive understanding of the interacting relations is gained. An optimal or near-optimal decision can then be made, such as desirable tolerance combinations, process settings, or reliability configurations.

Single Sampling. Sampling technique involving the sampling of a predetermined number of units, on the basis of which acceptability of a lot is evaluated.

Specification. Establishment of a desired or required goal, often involving a set of limits (upper and lower *specification limit*).

Specification Limit. See *tolerance.*

Standard. Goal value established as part of a *specification.*

Standard Deviation. Square root of *variance.*

Standard Error. Standard deviation of sample averages.

Standard Inspection List. A specification showing what quality characteristics are to be checked, how examination or measurement should be performed, and what limits or other values must be satisfied.

State of Nature. A condition regarding which we have probabilistic or else no knowledge, such as the percent defective in an uninspected lot.

Statistic. A *statistical measure.*

Statistical Measure. Value obtained as a result of mathematical operations on a sample, the *arithmetic mean* or *variance.*

Stress Analysis. Evaluation of electrical, thermal, vibration, shock, or other stress conditions under which components or equipments must function. Stress analysis results serve in predicting the failure rates to be expected in practice.

Subgroup. Usually, a subgroup is a *sample.* A subgroup is "rational" when the units comprising it are similar as to source, and therefore any variations among them may be viewed as not produced by an *assignable cause.*

Tolerance. Limit associated with a specification. For example, the specification: "No unit shall measure more than 1.003 nor less than 0.997 inch" sets tolerances of $1.000 \pm 0.003 = 0.997$ and 1.003 as lower and upper limits for each unit of product. The specification: "Average strength of ten skeins shall be 50 ± 2 kg." sets an upper tolerance of 52 and a lower tolerance of 48 kg. for the arithmetic average of a sample. An equivalent term is "specification limit."

Tolerance Failure. Equipment failure as a result of multiple drift and the interaction of instabilities within the system, even though no part failures have occurred.

Uncertainty. Situation in which no *prior probabilities* can be attached to a *state of nature,* such as whether or not a lot is of acceptable quality.

Unit. One of a number of similar items, objects, or other individual articles.

Universe. The totality of individual *units* under consideration. A day's, week's, or month's output of a particular item on a machine is an example of a universe or "population." The precision of all of a certain type of measuring instruments available in the marketplace would be the definition of another universe. The universe represents the aggregate from which the random *sample* for statistical analysis is chosen. When the *sample size* is less than 10 percent of the population, it is convenient (from a viewpoint of statistical analysis) to look on the universe as "infinite."

Useful Life. Total operating time between (1) debugging and shakeout and (2) wear-out.

Variability. General term denoting variation. More specific measures are the *variance, standard deviation, range,* and other *statistical measures.*

Variables Method. Measurement of quality of a unit of product along a scale, such as weight in pounds, resistance in ohms or moisture content in percent. An approach alternative to variable testing is *attributes* testing.

Variance. Average of the squares of the deviations of individual measurements from their average.

Variance Analysis. Statistical procedure for analyzing the sources of variation in a set of data.

Variation Coefficient. Standard deviation expressed as a percent of the arithmetic mean.

Wear-out Failure. Malfunction as a result of interaction of environmental stresses with factors producing equipment deterioration, such as abrasion, radiation, fatigue, and "creep," or corrosion and other chemical reactions.

Weibull Distribution. Failure distribution in terms of shape, scale, and location parameters. When the shape factor is greater than 1, a unimodal (one-peaked, bell-like) curve, though skewed, results. For a shape factor of 3.5, the form is close to normal. At 1.0, Weibull and negative exponential distribution are identical. When the shape factor is less than 1, the curve is steeper than the exponential form. The so-called Weibull probability density function, based on scale, shape, and location parameters a, b, and t respectively is $f(t) = abt^{b-1} exp(-at^b)$, from which it is apparent that when $b = 1$ the exponential function results.

Zero Defects Plan. A program of employe motivation, involving managers, foremen, and operators, seeking as a goal the elimination of errors, defects, and other failures that are humanly avoidable.

Index